Dilapidations and Service Charge Disputes

A Practical Guide

Simon Edwards, Keith Firn and Patrick Stell

2008

A division of Reed Business Information

Estates Gazette
1 Procter Street, London WC1V 6EU

ISBN 978-0-7282-0554-3

Cover design by Rebecca Caro
Typeset in Palatino 10/12 by Amy Boyle
Printed by Hobbs

Contents

Preface

Increasing case law and procedural rules governing commercial landlord and tenant relationships make it ever more important for clients, surveyors and lawyers to keep up to date on legal and technical issues relating to dilapidations and service charge disputes.

It is of equal importance that those actively involved in dilapidations and service charge disputes have a practical appreciation of how the legal and technical aspects combine; and how they can improve their chances of achieving a fair resolution to any dispute. This can be achieved by adopting just a few simple procedures.

This book offers clear guidance on all aspects of dilapidations and service charge disputes, in order to provide a practical understanding for any professional involved and their commercial property client.

By way of background, this book was initiated in 2007 at a time of uncertainty for the dilapidations professional in relation to the legal and practical aspects of dilapidations and service charge disputes.

The Royal Institution of Chartered Surveyors' (RICS) Service Charge Code had been published and practitioners were still adjusting to the Code's requirements that came into effect from April 2007.

The Property Litigation Association's (PLA) *Dilapidations Protocol* (version 2, dated September 2006) had been poorly received by surveyors in relation to a number of key aspects. The ensuing debate over the suitability of the PLA Dilapidations Protocol in turn delayed the fifth edition of the RICS *Dilapidations Guidance Note*. The PLA responded positively to the concerns raised and a revised *Dilapidations Protocol* (version 3) was released in May 2008. The revised RICS Dilapidations *Guidance Note* (5th Edition) followed in June 2008.

However, the "protocol" debate within the profession continues. It remains to be seen if the recently revised PLA Dilapidations Protocol will be formally adopted by the Lord Chancellor; or if the RICS' fifth edition *Dilapidations Guidance Note* used in conjunction with the Civil Procedures Rules (CPR) General Default Protocol provides more suitable and effective guidance.

At the RICS Dilapidations Forum annual conference in September 2007, the feedback gained from the delegates attending made it clear that surveyors, lawyers and landlord and tenant clients remained dissatisfied with the current dilapidation dispute resolution process, with an open recognition of the need for a more honest process. This book identifies that process and explains that dilapidations and service charge disputes have all they need in terms of statute, procedural law and common law to be negotiated effectively by properly qualified professionals.

In preparing this book, the authors have sought to provide practical guidance on how both commercial and service charge disputes sit within the current legal framework of England and Wales. They have also set out to provide practice guidance on dilapidations, looking at issues such as the concept and history of dilapidations and from taking instructions, through to resolving and settling a claim.

This book also looks at the typical service charge issues that both building owners/managing agents and occupiers face, such as dispute over the correct apportionment of the service charges, whether items that are recoverable via a service charge should be repaired or renewed (and who should pay, if replaced), as well as other problems and issues that arise.

The following chapters examine the common ground between the two areas of dispute; establishing the key differences; while providing pragmatic techniques to resolve conflict for the benefit of property professionals and their clients.

Case law, statutes and professional guidance are up to date as at June 2008.

Simon Edwards
Drivers Jonas LLP

Keith Firn
Barker Associates

Patrick Stell
GK Stell Building Consultants

Table of Cases

Table of Statutes

1 What are Dilapidations and Service Charge Disputes?

An introduction to dilapidations

A common question for the newcomer to the subject is "what are dilapidations?" A straightforward answer is not easy to offer. The concept is borne out of statute, common law and accepted procedure, formed over hundreds of years of such claims in England and Wales.

The nature of the legal system and the historical basis of property ownership in England and Wales over the centuries meant that the landlord's property asset was traditionally considered sacrosanct. Landlords leased property on terms designed to safeguard the property from any asset depreciation that would be caused by disrepair arising during the lease term.

Over the last 150 years or so, the emphasis placed on the landlord's rights has been eroded to such an extent that the concept of natural justice has flavoured statute and case law, to restore the balance or rights and liabilities between landlord and tenant.

The concept of dilapidations

There have been several attempts in the past to provide a succinct definition of dilapidations as a starting point for addressing the issue in its practical context.

The Property Litigation Association (PLA) Dilapidations Protocol (version 3, 2008) describes dilapidations as being "a claim for all breaches of covenant or obligation relating to the physical state of a demised property at the end of the lease".

However, this definition does not cover the broader range of issues present in many dilapidations disputes.

For example, the PLA description does not cover the common situation where there are reasonable grounds to commence a claim pre-lease end. In addition, the PLA's definition does not cover claims for damages under the law of waste, ie where there is no lease in place; where there are obligations to repair; and where the state of repair is in dispute between owner and occupier.

Perhaps due to the potential breadth of the topic, the latest Royal Institution of Chartered Surveyors (RICS) *Dilapidations Guidance Note* (5th ed, 2008) does not contain a simple definition of "dilapidations".

The definition of "dilapidations" used within this book is:

> A civil claim between landlord and tenant made under contract law and/or the law of waste where the claim is commenced during or at the end of the tenant's interest and where one party seeks a civil remedy and/or restitution from the other in respect of physical damage, deterioration or waste occasioned to or within the subject property that is alleged to be directly attributable to an act (or omission) by the other party.

The above definition is independent from the nature of the tenancy; applies during the term and at lease end; and does not restrict the concept to contract law.

What is a dilapidations dispute?

Dilapidations disputes have been described by some as a "black art" or as being the "sting in the tail" of commercial property leases. They are one of the few current forms of civil dispute where lawyers are not normally the principal professional advisors during the key pre-litigation stages of a claim.

In its most simple and common form, the dispute will be based on a schedule of dilapidations in which a claim for breach of the contract (ie breach of the lease) is prepared and then served by the landlord on the tenant.

The aim of the schedule is to document the individual breaches and to set out works required to remedy each breach, together with associated costs. The tenant is required to either carry out the work or pay damages for the breaches at lease end.

The calculation of any damages due to the landlord is the most contested area of any dilapidations dispute. To resolve disputes effectively, technical and legal considerations must be factored in to clearly quantify the liability or "loss" suffered by the landlord.

Dilapidations disputes are thus a rare hybrid of surveying and legal issues, where the lawyer must take a keen interest in the role of the surveyor and vice versa. This is where the chartered building surveyor, the valuer and the property lawyer earn their keep.

Why dilapidations disputes occur

There are a multitude of reasons why dilapidations and service charge disputes arise between landlord and tenant parties:

- It may be as simple as one of the parties having been naive in agreeing lease terms and, as a consequence, not having taken appropriate advice on their future liabilities before committing to the lease.
- There may be financial difficulties that would make it problematic for a party to honour tenancy obligations.
- It could be that allowing disrepair to occur is part of a deliberate property management strategy that arises from a tenant's cost-to-risk appraisal.
- It could be that there has been an unfortunate breakdown in relations and/or communication between the parties.

The possible causes of dilapidations and service charge disputes are extensive. Wherever disputes arise there will be a dispute resolution role for surveyors and other advisers. This may require the surveyors and other advisers to first review, determine or clarify their client's position; and then to act to resolve the dispute while safeguarding their clients' interests to the best of their ability, without acting unreasonably or even unlawfully in the process. There are a multitude of reasons why dilapidations and service charge disputes arise between landlord and tenant parties.

Statute and common law influences

Dilapidations disputes sit within the legal framework of statute and common law. Basic knowledge of both areas of law is required in order to accurately resolve claims.

Statutes are acts of Parliament that may date back centuries. There are many examples of statutes relevant to dilapidations and examples of key current legislation include: the Law of Property Act 1925; the Landlord and Tenant Acts of 1927 and 1954; the Leasehold Property (Repairs) Act 1938; and the Defective Premises Act 1972.

Detailed interpretation of statutes relating to dilapidations is provided by "common law", ie legal opinion embodied in a judicial decision. Dilapidations common law also stretches back hundreds of years, providing a valuable insight into the equitable remedies imposed by the courts for a wide variety of commonly encountered claims and disputes.

Perhaps the most well known dilapidations case is *Proudfoot* v *Hart* (1890) LR 25 QBD 42 that clarifies the nature of the tenant's obligation to repair. The judgment in this case set down that the determination of whether a building is in actionable disrepair at lease end has to take into consideration the age, character, and location of the premises and whether it would be "reasonably fit for the occupation of a reasonably minded tenant of the class who could be expected to take it at the start of the lease". This case remains relevant today and is widely quoted.

Common types of schedule

Interim schedule

An "interim schedule" is the term commonly used by surveyors and solicitors to signify that a schedule of dilapidations that is served during the currency of a lease where there is three years or more remaining in the lease term.

The origins of the "more than three years to go" timeframe can be found in the Leasehold Property (Repairs) Act 1938 (LP(R)A 1938) that entitles qualifying tenants to claim the benefit of the act and gain relief from forfeiture or re-entry proceedings commenced under section 146 of the Law of Property Act 1925 (LPA 1925).

The LP(R)A 1938 may apply if there are three or more years to run in a lease term and so interim schedules evolved to distinguish them from schedules and claims made in the last three years of a lease where the LP(R)A 1938 does not apply.

Terminal schedules

A "terminal schedule" is the term commonly used by surveyors and solicitors to signify a schedule served during the last three years of the currency of a lease; or at the end of the lease (see also final or lease end schedules).

The origins of the "last three years of the term" timeframe is again associated with the LP(R)A 1938. Tenants in the last three years of a lease will not have any right to seek relief from forfeiture or re-entry proceedings under section 146 of the LPA 1925.

Lease end or final schedules

Some surveyors use terms such as "lease end" or "final" schedules as a means of distinguishing schedules prepared after a lease has irrevocably determined from a "terminal schedule" previously served during the currency of the lease.

The choice of label for post lease end schedules is a matter of semantics and personal preference. However, the use of the word "final" in schedule descriptions can cause confusion if a final schedule is subsequently revised and re-issued (possibly more than once).

Format of the schedule of dilapidations

With a schedule, the typical claim is broken down into sections or areas of claim covering different types of breaches of the lease, such as wants of repair; redecoration; reinstatement of alterations, etc. The claim may also set out claims for consequential losses arising from the breaches, such as loss of rent, fees, VAT, etc.

A basic lease end schedule will, typically, be similar in format to the following example.

Item	Lease Clause	Breach of Covenant	Remedy	Cost
1	3.42	The tenant has installed 5 l/m of base units and 4 l/m of wall units to form a kitchenette	Strip out base and wall units, remove from site, make good	£325
2	3.42	The Tenant has tiled the splash-back area above the work top (5 l/m)	Remove tiling. Make good plastered surface ready for redecoration	£285

Typical lease end schedule

However, it is becoming increasingly common for schedules to be prepared and served from the outset in a "Scott" schedule" format (see Chapter 11).

Which type of schedule of dilapidations to serve?

The choice of the type of schedule will depend on the nature of the remedy being sought and the length of unexpired term of the tenancy. Commonly encountered schedules include:

- Interim or terminal schedules — or service under cover of a LPA 1925 s146 notice.
- Interim or terminal schedules — for service under cover of a "notice to repair".
- Terminal or "lease end" schedules — for service seeking damages following the end of a lease or tenancy.
- Lease "break option" schedules.
- Lease assignment schedule.

Life cycle of a typical claim

Increasingly, landlords and tenants of commercial property are aware of their contractual rights and potential defences. It is therefore unsurprising that the typical commercial landlord and tenant will view such a liability as a contentious matter. Dilapidations claims generally take the following course:

1. Preparation of a schedule by the landlord's appointed surveyor.
2. Service by a lawyer.
3. The tenant's response in a "Scott" schedule format.
4. (a) Tenant does the work before the end of the lease; or
   (b) negotiates and pays damages (where applicable).
   (c) Litigation — if steps 4a or 4b are unsuccessful in resolving the dispute, the surveyor(s) will be formally appointed as an "expert witness".

Resolution through litigation

Only a very small percentage of disputed claims are litigated, when the claimant (typically the landlord) commences court proceedings against the defendant (typically the tenant).

These claims litigate for several reasons, including:

- intransigence of the client/surveyor
- complicated and/or ambiguous technical or legal issues.

Court action to resolve a dilapidations dispute is always considered as an option of last resort due to the litigation costs. Other less costly methods of alternative dispute resolution (ADR) should also be considered before litigation is commenced.

Where litigation appears the only option to resolve a dispute, it is commenced when the claimant (normally the landlord) prepares the documentation ready to issue proceedings in the court. Once the court documents and evidence are ready, the claimant pays the court fee (typically a few hundred pounds) and the claim then starts the sometimes slow procedure towards a final court hearing.

Court proceedings are normally concluded by a court order, which can be either based on agreement by the parties or on a final judicial decision. At the point of termination of the court proceedings, an award of costs in favour of the successful party is generally made, unless otherwise agreed between the parties. These costs can include legal costs, surveyor's fees, and other expenditure related to the claim. The risk of an adverse costs award is often considered to be a significant risk when starting court proceedings and a significant incentive to settle the dispute early.

An introduction to service charges

The concept of a service charge

Service charge is the term commonly used to describe the collective running costs of a shared property or estate that each individual occupier or tenant is obliged to contribute towards. The scope of services that are to be undertaken by the landlord and recovered from the occupiers are agreed before the lease commences and form part of the lease contractual terms. Often, the tenant's percentages contribution towards the total services charges are also defined by reference to a fair proportion; a set percentage or formulae for contributions; and/or a maximum sum of the estate service charge total.

If there are no terms within a lease requiring a tenant to pay a service charge, then the landlord cannot introduce one at a later date without the tenant's agreement (which they are unlikely to get). See *O'May* v *City of London Real Property Co Ltd* [1982] 1 EGLR 76, referred to in Chapter 16, for further reference to this.

Statute definitions of service charges

There is no statutory definition of service charges in relation to commercial premises; but there is a statutory definition of services charges for residential premises. Section 18 of the Landlord & Tenant Act 1985 (LTA 1985) states that residential service charges are:

> an amount payable by a tenant of a dwelling (removed for commercial leases) as part of or in addition to the rent —
> (a) which is payable directly or indirectly for services, repairs, maintenance or insurance or the landlord's costs of management, and
> (b) the whole or part of which varies or may vary according to the relevant costs.

If a lease, whether of commercial or residential premises (or a mixed-use scheme) states the above phrasing, or makes reference to section 18 of the LTA 1985, then it is likely that a landlord should be able to recover as much "service charge" expenditure as possible from the tenant or tenants of a building.

Although this wording is only for residential premises and not commercial, it is often used to define what is, and what is not, a service charge in commercial leases; the only difference being the omission of the words "of a dwelling" in the section.

Why service charge disputes occur

Service charges are one of the major recurring costs that businesses face, and are sometimes closely related to dilapidations claims.

In most leases, there will be degree of certainty over ongoing tenant expenditure. For example, rents are typically fixed until the next review and business rates are capped to any annual phasing rules.

However, it remains possible for landlords to dramatically increase their service charge budget from year to year and this in turn can cause havoc to tenants' budgeting and profitability.

Unless tenants have a cap on their service charges or have exclusions on their liability set out in their leases, they cannot predict what the costs will be from one year to another. Therefore, it is no surprise that disputes sometimes arise between landlords and tenants as to the sums that tenants are being asked to pay.

Often these disputes are about whether items such as lifts or air-conditioning systems can be repaired or whether they should be replaced. That type of dispute is where the tenants may be seeking a contribution from the landlord towards the costs of replacement of "big ticket" items of expenditure.

Another type of service charge dispute is not between the landlord and the tenants, but between the tenants themselves. Tenants may feel that other occupiers of an office building or shopping centre should pay a higher proportion than the proportion that the landlord has assessed them to bear.

When this occurs in an office building or a shopping centre, the landlord will usually look to the amount of floor space each tenant occupies and assess their service charge contributions based on that. However, the amount of floor area occupied by each tenant may not always produce a satisfactory solution to the amount of service charge each tenant should pay. An example of a type of property where this often will not be appropriate is that of a mixed-use leisure scheme.

This book intends to look into these types of disputes in turn and give advice on how to deal with the problems.

Service charges and dilapidations

During the term, there is rarely a direct relationship between service charge payments and dilapidations liability. Service charge payments are for works that have been arranged to be carried out by, or on behalf of, the landlord and the recovery of those costs is normally during the financial year that the costs are incurred. Whereas the tenant's lease end dilapidations liability accrues over a longer period of time and that liability only manifests itself when, or if, the landlord serves a schedule of dilapidations on the tenant.

In accounting terms, the tenant will wish to budget for the service charge items that will need to be paid during that financial year. The tenant will probably not be aware of the costs that would need to be budgeted in order to meet any likely dilapidations payments at the lease end.

As stated above, there is not a direct relationship between the service charges and dilapidations during the term of a lease. However, where a full repairing lease of a whole building has led to a number of subleases being created from it, there is a direct relationship between service charge payments made during each financial year of the building's service charge accounts and the dilapidations liability. This is because the original tenant whom had those full repairing liabilities on all of the plant and equipment, the roof, windows, etc is now the "reluctant landlord".

Where a property has been sublet (possibly to multiple sub-tenants), the original or "head" tenant or landlord will probably wish to mitigate their dilapidations liability on all of the plant and equipment and other items in the common parts shared by the tenants or sub-tenants by retaining the lease obligation to repair and maintain the common items on a service charge basis to the sub-tenants.

The head tenant or landlord will be fully aware of their own liability to have the whole building repaired and with the plant and equipment in good order at the end of the lease(s). By retaining liability for common parts and services, etc and by implementing a planned maintenance programme of maintenance, repair, cyclical decorations or even renewal works, the head tenant or landlord can effectively keep these common items in full and good repair and condition at the (sub)tenant's expense during the terms of their leases without having to repeatedly commence difficult dilapidations proceedings throughout the term. If carefully managed and operated, the head tenants or landlords will keep their own lease end liability for dilapidations as low as possible.

2 Dilapidations, Statute and Common Law

Part 1: The history of dilapidations

Dilapidations claims are considered to be complex due to many hundreds of years of statute and common law judgments, each one building on existing principles. To understand how modern dilapidations processes have evolved, it helps to have an appreciation of the historical origins of this unique area of dispute.

For as long as people have owned property which they allowed others to occupy, there have been claims between disgruntled parties relating to damage, disrepair and deterioration of the fabric of the building during the period of third-party occupancy and at the end of the lease. How these claims were dealt with in the pre-early medieval period is not readily apparent, due in part to the lack of surviving records. When seeking to identify the beginnings of the concept of "dilapidations" in England and Wales, the earliest point of reliable reference are records and statutes dating from the 13th century.

13th century — the medieval origins of dilapidations?

Arguably, the origins of modern dilapidations can be traced back to a collection of acts passed by Henry III within the Statute of Marlborough in 1267, one of which, The Waste Act 1267, granted the right to make claims for damages for "waste" subject to there being no "special licence ... (by) covenant ... that they may do it". (For example, no contractual relationship or covenant permitting an act of waste to occur).

At the time of the enactment, waste was the name given to the concept of damage arising from a direct or positive act by one party that caused damage to property (including land, buildings, trees, etc). The successful claimant would be able to obtain a single measure damage for waste incurred. This statute remains on the statute books and remains the cornerstone for modern claims for dilapidations under the law of waste. In modern terms we would call this damage for "voluntary waste".

An early problem arose with the law, that while a claimant could gain single measure damages for waste, they were not entitled to their costs incurred in the action. Consequently, Edward I passed the acts collectively known as the "first" Statute of Westminster 1275, a result of which the claimant became entitled to also recover the cost of the litigation in limited circumstances.

In 1278, the Statute of Gloucester was passed which included the Action for Waste Act 1278; and the Recovery for Damages and Costs Act 1278. The latter allowed a claimant to seek treble damages for waste from a "tenant for life" or a "tenant of years", ie a punitive remedy that remained on the statute books until repealed by the Civil Procedures Acts Appeal Act 1879.

In addition to laying the foundations for modern dilapidations claims, Edward I also gave royal assent to the Statute of Quia Emptores 1290, which contained the Restraint of Subinfeudation Act 1290. This act was aimed at addressing the many uncontrolled assignments of interests in land that had and were taking place since the Doomsday Book in 1086. The reason for Crown intervention was that it was becoming impossible for the Crown to keep track of who owned the land and, more importantly, what taxes were due. Such was the significance of this act that some historians argue that the Statute of Quia Emptores was the land ownership platform on which the Houses of York and Lancaster built their power bases, a situation that ultimately resulted in the War of the Roses and a change in the English royal dynasty 195 years later in 1485.

14th to 16th centuries — leasehold developments

Concerns relating to costs were further addressed with the introduction of the Untrue Suggestions in Chancery Act 1393. The statute gave the Lord Chancellor discretionary power to award damages to a defendant who successfully defended a claim. This power was further confirmed and extended in 1436 to provide a new power to take or seek security for costs.

Throughout the 14th and 15th centuries, claims continued to be made for waste, but the issue of recovery of costs and damages in waste claims continued to prove difficult.

The 16th century saw a number of developments in landlord and tenant statute that sought to develop the contractual nature of a lease and to make clearer the rights, benefits and obligations of the lease parties.

Following the dissolution of the monasteries by Henry VIII between 1536 and 1541, the Crown became the owner of vast tracts of previously church-owned land and property. With these widespread changes, new laws were required to help administer, manage and protect the extended crown estates. Henry VIII introduced the Leases Act 1540; the Leases by Corporation Act 1541, the Repair of Decayed Houses, England and Wales Act 1543 and the Leases Act 1544. Later, his daughter Elizabeth I, introduced three Ecclesiastical Leases Acts (1571, 1572 and 1575) and a Universities and Colleges (leases, etc) Act in 1575.

The spate of acts passed in the mid to late 16th century dealing with commercial ("corporation"), ecclesiastical and educational property interests and leases, promoted better and more widespread landlord and tenant contractual relations and started to move the emphasis away from law of waste claims to contractual-based claims.

17th century — cost and contractual damages

At the start of the 17th century, James I (and IV) introduced the Costs Act 1606, which gave defendants the ability to recover costs where they were successful in their defence of a claim (now a fundamental legal principle). It appears that there was also widespread concern with claims for costs being abused and a perception of unreasonable professional fees. To address this perception, James I introduced the Attorneys Act in 1605 which sought to curb professional fees and costs incurred in actions.

By the mid 17th century, common law judgments in waste claims had also extended the concept to include a right to claim damages from "tenants for years" for what we now call permissive waste, ie waste and damage arising as a consequence of a failure or neglect to act. Claims for permissive waste however, were not permitted against any classes of tenancy other than a "tenant for years" and this principle remains today.

By 1670, the ability to recover damages under contract law had become preferable to action for waste under the Statute of Marlborough. It was later observed by Sergeant Williams commenting on the case of *Green* v *Cole* (1670) that the action for waste "is now very seldom brought, and has given way to a much more experienced and easy (alternative) remedy by 'an action on the case'".

He went on to observe, "this ... action (for waste) was found by experience to be so imperfect and defective a way of recovering seisin of the place wasted that the plaintive attained little or no advantage from it ..."

When commenting on dilapidation actions, Sergeant Williams further noted:

> where the demise was by deed, care was taken to give the lesser a power of re-entry, in case the lessee committed any waste or destructions; and an action on the case was then found to be much better adapted for the recovery of mere damages ... It also has the further advantage that it may be brought by him in the reversion or the remainder for the life of years as well as in fee, or entail; and the plaintive is entitled to costs in this action which he cannot have in an action for waste.

Clearly, by 1670, suing for contractual dilapidations damages had become the more normal course of action. Perhaps coincidentally, in 1670, the Duties on Law Proceedings Act 1670 was passed that limited the amount of costs that could be recovered by a claimant in a personal action where damages of under 40 shillings (£2) were awarded. Modern costs judges have suggested that this act was effectively the equivalent of the small claims court fast-track route available today.

The introduction of the Statute of Frauds 1677 created the requirement that certain kinds of contracts, such as the transfer of an interest in land, be made in writing and that they are signed. While most of the 1677 Act has been repealed, the legal principle for the transfer of an interest in land to be in writing remains a basic legal principle re-enacted in modern statute.

Interestingly, The Private Act 1697 was a private Bill/Act passed in 1697 which cleared the Bishop of Ely of his "dilapidations" liability, highlighting the fact that the phrase dilapidations was by this stage generally associated with ecclesiastical disrepair and chancery repair claims. In commercial matters, similar contractual-based dilapidations damages claims continued to be commonly referred to as "an action on the case".

18th century — repair and landlord and tenant developments

During the 18th century, there were further statutory developments with the introduction of the Landlord and Tenant Acts in 1709 and 1730; the Clergy Residencies Repair Act 1776; the Ecclesiastical Leases Act 1765 and the Clergy Residencies Repair Act 1781. These acts show continued statute interest and development in the concept of leasehold repairs, but had minimal impact on the procedural laws affecting commercial dilapidations.

19th century law reform

Between 1770 and 1830, the dominant parliamentary force was the Conservative Party. During this period, they introduced the Ecclesiastical Leases Act 1800 and the Recovery of Possessions by Landlords Act 1820. By the 1820s and 1830s however, the Conservative government had become unpopular and there was substantial unrest with riots on the streets and a threat of revolution. The general malcontent led to demands for legal and statutory reform such that the Conservative government was swept away and replaced with the Whig administration led by Lord Grey, who formed a government based on modernisation and a platform of reform. The Reform Act 1832 heralded widespread statutory reform.

At the time, the means by which land in England and Wales was conveyed was still based on the archaic feudal laws dating back to the 13th century. These were changed, first by the Fines and Recoveries Act 1833 and then the Conveyance by Release Without Lease Act 1841. Further significant property law changes occurred with the passing of the Law of Real Property Act 1845, the Conveyance of Real Property Act 1845 and the (Granting) Leases Act 1845.

From a dilapidations perspective, the (Granting) Leases Act 1845 created a model commercial lease, similar in nature to the current institutional leases or Law Society short form lease. However, it was not mandatory to use the model lease and parties remained free to agree their own alternative contractual lease terms.

The reform continued apace and two further lease(s) acts followed in 1849; together with the Renewable Leasehold Conversion Act 1849. Another leases act was passed in 1850 and a Landlord and Tenant Act in 1851. There were two Law of Property Amendment Acts in 1859 and

1860; Ecclesiastical Dilapidations Acts in 1871 and 1872; and a Tenants Compensation Act in 1890. Statute concerning conveyance of property was also significantly simplified and amended, with five conveyancing-related acts between 1874 and 1911.

20th to 21st century — the modern framework

The volume of statute passed in the mid to late 19th century were fast becoming unwieldy and created a need for further rationalisation and simplification. In the 1920s, a programme of reform was initiated with three separate Law of Property Acts passed between 1922 and 1925; the latter of which remains largely in force to date.

Since 1925, the pace of statutory changes and reform have continued, with principal legislation being the Landlord & Tenants Acts 1927 and 1954; the Leasehold Property (Repairs) Act 1938, and various other subsequent landlord and tenant or leasehold-related legislation that makes up the dilapidations statutory framework today.

Summary

The concept of dilapidations-focused statutes in England and Wales can be traced back over 750 years to the Statute of Marlborough 1267. In terms of dilapidations common law, the vast majority of dilapidations statutes and precedents that affect modern claims date from after 1850. The volume of modern case law reflects the substantial statutory reform since 1845.

While most modern dilapidations disputes are generally initiated as claims for contractual damages, the more ancient basis for a claim under the law of waste continues to be of use in cases such as *Dayani v Bromley London Borough Council* [1999] 3 EGLR 144.

Part 2: Modern dilapidations statutes

The ability to pursue a dilapidations claim is subject to an extensive body of statutes, regulations, and byelaws concerning the law of property, landlord and tenant law, health and safety law, and employer and occupier law to name but a few.

All relevant modern dilapidations statute and subservient legislation cannot be covered in detail in a general book on the subject, as it could easily warrant an entire book on the topic. Therefore, we have assumed that our readers will have access to modern IT/web-based sources, via which copies of the acts and detailed guidance can be obtained when required. For example, copies of Acts of Parliament may be purchased from the Stationery Office Limited (see *www.tsoshop.co.uk*); or where available can be found on the Office of Public Sector information website at *www.opsi.gov.uk/acts*.

If the latest versions of Acts of Parliament and subservient legislation complete with modifications, amendments and revisions, etc is required, then the Ministry of Justice website at *www.statutelaw.gov.uk* has an excellent advance search facility which will let you view amended acts, etc. This website will also let you check to see if there have been recent changes made to the subject statute, such as repeals or amendments that have yet to be incorporated into the online version of any act, which will ensure that you work within the latest legal framework at all times.

While we do not intend to repeat and comment on the extensive modern statute in its entirety, a brief summary of key dilapidations-related statute and subservient legislation is set out below.

Law of Property Act 1925

From a dilapidations perspective, surveyors engaged in dilapidations claims should have a general working knowledge of Part V of the Law of Property Act 1925 (LPA 1925). In particular, surveyors should be familiar with the requirements of the following sections of the Act.

Section 136 (Legal assignments of things in action)

By virtue of section 136, two parties may come to an agreement in writing, whereby the right to bring a claim such as a "debt or thing in action" can be transferred or "assigned" to a third party not part of the original contract relations. There are wide definitions of types of action that can be assigned, including the right to make a claim for dilapidations.

It is relatively rare to come across cases where the rights to claim dilapidations have been assigned, but surveyors should be aware of this possibility and obtain advice on the implications of an assignment where necessary.

Section 146 (Restrictions on and relief against forfeiture of leases and under leases)

Section 146 is commonly encountered by a surveyor when preparing or responding to a dilapidations claim where the intended remedy is for forfeiture, re-entry or damages during the lease term. If the action is to be successfully implemented, then great care should be taken to follow the notice and action procedures set out in section 146; and in particular to ensure that any notices for forfeiture are compliant with section 146(1).

Regard must also be had for the requirement to allow the tenant to be afforded a reasonable period to undertake the works (see section 18(2) of the Landlord and Tenant Act 1927 (L&TA 1927). Where relevant, the landlord must notify the tenant of their right to seek relief from forfeiture or re-entry under section 1 of the Leasehold Property (Repairs) (LP(R)A) Act 1938 (see below) and/or from the courts under section 146(2). Where relief is sought, a landlord would be entitled to recover their costs under section 146(3).

Section 147 (Relief against notice to affect decorative repairs)

Under section 147, where a landlord serves any notice that includes a request to undertake internal decorative repairs during the lease term, the tenant may seek relief from the court either wholly or partially during the lease term, subject to criteria contained within the section.

Section 148 (Waiver of a covenant in a lease)

Surveyors acting on dilapidations may, on occasion, come across circumstances where an aspect of a claim is challenged on the basis that there has been some alleged "waiver" of rights and, in such circumstances, should seek specialist legal advice on the impact of section 148 of the Act.

Section 150 (Surrender of a lease without prejudice to under leases with a view to granting a new lease)

Dilapidations surveyors may also, on occasion, need to take specialist advice on the application of section 150 in the circumstances where a landlord is seeking a surrender remedy against a tenant where there is a sub-tenant remaining within the whole (or part) of the property.

Section 196 (Service of notices)

Section 196 is relevant for the service of notices such as section 146 notices seeking forfeiture, etc. It is important that the surveyor involved in the dilapidations proceeding is familiar with the law and service of such notices under section 196, as there may be circumstances where an invalid or inappropriately served notice may have significant consequences for a claim. In certain circumstances, this can invalidate parts or even the whole of the claim or prevent a remedy such as forfeiture or re-entry from being successfully implemented. Great care must be taken when assisting in the preparation of the notices; or receiving and reviewing notices to ensure that they have been served in accordance with the appropriate lease obligations and statute.

Landlord and Tenant Act 1927

Part 1 — sections 1 to 17 (Compensation for improvements, etc)

Part 1 (sections 1 to 17) of the Landlord and Tenant Act 1927 (L&TA 1927) contains procedures and rights available to tenants for claiming compensation for improvements undertaken by the tenant (or their predecessor) during the lease term. This part of the Act is often misunderstood and overlooked in dilapidations claims as it is somewhat complex to follow due to the extensive amendments made by Part III, sections 47 to 50 of the Landlord and Tenant Act 1954.

At its simplest (and subject to following the strict statutory procedures), at lease end a tenant can claim for compensation for "improvements" (including improvements undertaken in order to comply with statutory obligations) where any compensation due can be offset against any landlord's dilapidations damages claim.

Where surveyors are engaged in giving strategic dilapidations advice to tenants during the currency of the lease, the issue of licences to alter and compliance with the procedures contained in Part I of the L&TA 1927 (as amended) may prove invaluable for mitigating a tenant client's end-of-term dilapidations liability.

Section 18(1) First limb (Provisions as to covenants to repair)

Dilapidations surveyors most commonly come across section 18(1) when considering the extent of financial damage to a landlord's reversionary interest arising from breaches of repair covenants (which the courts have confirmed also includes aspects of "decorative repair" — see *Latimer* v *Carney* [2006] 3 EGLR 13).

Section 18(1) is a key piece of statute and is described as having two limbs. The first limb states:

> damages where a breach of covenant or agreement to keep or put premises in repair during the currency of the lease, or to leave or put premises in repair at the termination of a lease, where such covenant or agreement is expressed or implied, and whether or specific, shall in no case exceed the amount (if any) by which the value of the reversion (whether immediate or not) in the premises is diminished owing to the breach of such a covenant or agreement as aforesaid; ...

In layman's terms, this can be summarised as meaning that the amount of recoverable damages associated with repair or decorative repair works necessary to remedy a breach will be capped to the damage to the reversionary interests for the property (the value of the property), ie the wronged party can only recover what it has actually lost.

Before surveyors complete the preparation of dilapidations damages claim or response, careful consideration must be given to whether or not there has been any damage to the reversionary interest arising from breaches of repair or decorative repair covenants; and if damage has occurred, whether it exceeds the damage caused to landlord's reversionary interests.

Section 18(1) Second limb

The second limb of section 18(1) states:

> and in particular no damage will be recovered for a breach of any such covenant or agreement to leave or put the premises in repair at the termination of the lease, if it is shown that the premises, in whatever state of repair there might be, would at or shortly after the determination of the tenancy have been pulled down, or structural alterations made therein as would render valueless the repairs covered by the covenant or agreement.

In layman's terms, the second limb only applies to terminal or lease end claims and will impose a cap on the level of recoverable damages where any remedial works, had they been undertaken by the tenant, would have been "rendered valueless" (often referred to by surveyors as "supersession"). For example, a landlord's claim for the costs of painting a wall where the landlord intends to demolish the wall regardless of decorative condition would be rendered valueless as the decorative state of the wall would be of no material relevance as it would be demolished in any event.

However, even where alternative works are planned that may result in material changes that, on the face of it, may render a claim valueless, it remains possible for parts of a claim to "survive" when the sequence of both the dilapidations remedial works and alternative works intentions are considered and compared in sequence.

Surveyors should always seek to consider both supersession and survival carefully.

Section 18(2) (Extending section 146 of the LPA 1925)

Section 18(2) extended section 146 of the LPA 1925 (see above) so that a person serving a section 146 notice can't enforce the notice and action unless they can prove that the notice was adequately served on the tenant/recipient (see section 196 of the LPA 1925); and that the notice recipient then had a "reasonable" period after service in which to undertake the requested remedial works.

Section 19 (Provisions as to covenants not to assign, underlet, part with possession without licence or consent)

Section 19 (where it still applies) contains provisions as to covenants not to assign without licence or consent. Section 19(2) may be of particular relevance when considering issues of tenants alterations and is worthy of further consideration by surveyors if disputes arise during dilapidations proceedings about tenants' unauthorised alterations within a property.

Section 23 (Service of notices)

Section 23 concerns the service of notices under the Act and is of particular relevance for consideration where claims for compensation for improvements are made by a tenant under Part 1 of the Act (see above).

Leasehold Property (Repairs) Act 1938

The Leasehold Property (Repairs) Act 1938 (LP(R)A 1938) was introduced to correct a perceived "mischief" that was prevalent at the time following fairly widescale abuse of the landlord's right for forfeiture on re-entry contained in section 146 of the LPA 1925. The Act sought to introduce safeguards in the tenant's favour to prevent exploitative landlords from unreasonably seeking forfeiture for trivial and relatively minor wants of repair.

Section 1 (Restriction on enforcement of repairing covenants in long leases)

Section 1 imposes a restriction on enforcement of repairing covenants whereby tenants in receipt of a section 146 notice may claim the benefit of the Act and relief from forfeiture or re-entry. Where tenants claim the benefit of the Act and the landlord wishes to proceed with forfeiture, the landlord will need to gain leave of the courts to do so and the courts may impose the appropriate terms. The grounds for seeking leave of the courts are expressly stated in section (5)(a) to (e) of the Act. The LP(R)A 1938 also imposes restrictions on the landlord's right to recover costs where the tenant successfully claims benefit of the Act.

The LP(R)A 1938 is not applicable in all circumstances and in particular, business tenancies cannot claim the benefit of the Act if the lease is for less than seven years of term; or if the lease is for a longer period but there is less than three years remaining of the term.

There are other circumstances where the tenant's right to claim benefit of the Act does not apply. It should also be noted that the original Act related to houses but was subsequently extended to apply to all types of property and was further varied in terms of application by way of section 51 of the Landlord and Tenant Act 1954.

Because the rights, benefits and application of the Act are often misunderstood or incorrectly applied, any surveyor acting on a dilapidations claim during the currency of a tenancy should give careful consideration to the LP(R)A 1938 (see the flowchart in Chapter 6) and should advise their clients accordingly of the tenant's potential rights to claim benefit under the Act.

Landlord and Tenant Act 1954

The Landlord and Tenant Act 1954 (L&TA 1954) introduced wide-ranging revisions to the law of landlord and tenant which particularly sought to revise and clarify procedures previously introduced under the L&TA 1927 (see above).

From a dilapidations perspective, surveyors engaged in dilapidations should be generally familiar with sections 23 to 46 contained in Part II. A dilapidations surveyor, during the course of a claim, may need to consider issues arising under the following sections of the Act:

Section 25 (Termination of tenancy by the landlord)

If the landlord wishes to terminate the lease during the term or at the lease end, then they will need to serve the tenant with a section 25 notice of their intentions. Depending on the circumstances that gave rise to the notice, the tenant may be entitled to seek "compensation" from the landlord. Any compensation due may be offset against any dilapidations damages otherwise due to the landlord.

Where a section 25 notice has been served on the tenant, the surveyor should seek a copy as it may contain materially relevant information that effects or restricts the measure of damages for loss in the landlord's dilapidations claim.

Section 26 (Tenant's request for a new tenancy)

Where a tenant requests a new tenancy under section 26, it is often considered to be reasonable grounds to refuse a tenancy where there are unresolved tenant dilapidations issues. Landlords may alternatively require any dilapidations works to be undertaken within a set period at the commencement of a new lease within an expressly specified period (and if so the tenant will not be able to claim the benefit of the LP(R)A 1938 if they then fail to undertake the works and the landlord seeks forfeiture).

Where a section 26 notice has been served by the tenant, the surveyor should seek a copy (and a copy of any landlord's response) as it may contain materially relevant information that affects the landlord's lease end dilapidations claim or dilapidations strategy under the new lease.

Sections 24 to 28 and section 38 (Provisions for tenants to "hold over")

The dilapidations surveyor may need to review the lease and establish whether or not a tenant apparently "holding over" within a property beyond the end of the original lease term is doing so as a continuation of the lease; or whether they are "1954 Act excluded" by way of a court order with no rights to hold over. In the latter case, any continued occupation would be in the capacity as a tenant at sufferance or a tenant at will.

Where a tenant at sufferance or a tenant at will tenancy exists, the dilapidations surveyor will need to consider the remedies of any breaches during the original lease term in the normal manner, but may be restricted to claims under the law of waste for any disrepair or waste arising at the property after the date the act excluded lease determined.

Section 40 (Duty of tenants and landlords of business premises to give information to each other)

Section 40 imposes a duty on the landlord and tenant to give information to each other concerning their tenancy interest and identities of parties etc within one month of the date of the notice.

It is not uncommon during dilapidations proceedings to find evidence that suggests that there has been an unauthorised under-letting or parting with possession of the property; and that there may be a "tenant at estoppel", "tenant at will" or "tenant at sufferance" in the property without landlord knowledge or consent.

Where such issues are identified, the landlord's solicitor should be able to make an appropriate request under section 40 to gain the information from the tenant as to who is in the property so the dilapidations surveyor and the landlord can further consider the appropriate remedies to be implemented.

Sections 47 to 50 (Compensation for improvements)

Under sections 47 to 50, modifications and revisions were made to Part 1 of the L&TA 1927 concerning a tenant's right to seek compensation improvements (see above). Section 47 made revisions as to the procedures/timing for a tenant seeking to claim compensation for improvements. Section 48 made amendments as to the limitations on the tenant's right to compensation. Section 49 introduced restrictions on contracting out of the rights for compensation for improvements. Section 50 contained general interpretation guidance.

Section 51 (Extension of Leasehold Property (Repairs) Act 1938)

Section 51 introduced extensions to the LP(R)A 1938 which widened the application of the 1938 Act to different property types (other than houses). It also varied other aspects of the 1938 Act as to when it applied in relation to particular leases according to the duration of the tenancy of the lease and the term remaining unexpired (see the flowchart in Chapter 6).

Occupier's Liability Acts 1957 and 1984

The Occupier's Liability Act 1957 (OLA 1957) imposed a "common duty of care" towards visitors to occupied premises.

The duty of care requires the occupier to manage risks to safeguard visitors to a property. Subsequent case law has found occupiers liable for damages where defects and disrepair to properties and structure pose a safety hazard that caused an injury.

From a dilapidations perspective, the common duty of care should be taken into consideration when surveying buildings and property, as it may provide an opportunity for the surveyor to further demonstrate the reasonableness of any claim being made on the tenant with regards to disrepair and deterioration at the property which poses a hazard.

It is also of assistance in considering the extent of breaches of compliance with statute, etc, and also the extent of a tenant's compliance with the implied obligation of "tenant-like user".

The common duty of care under the OLA 1957 applies only to lawful visitors to a property.

However, the Occupier's Liability Act 1984 imposed similar civil law requirements on the owner/occupiers to protect those visitors who are not there lawfully, including trespassers.

Defective Premises Act 1972

From a dilapidations perspective, the key section of the Defective Premises Act 1972 (DPA 1972) is contained within section 4 of the Act.

Section 4 (Landlord's duty of care in virtue of obligation or right to repair premises demised)

Under section 4, the landlord can gain a duty of care for damage or injury caused by disrepair at a property, even if the tenant has an obligation to repair the property under the lease.

To summarise section 4 obligations, if the landlord becomes aware of a defect at a property for which the tenant is liable, then the landlord owes a duty of care to take all reasonable action to protect people who could be exposed to the risk of personal injury posed by the defect. As a consequence, should a defect be identified during a dilapidations inspection then it would be considered that the landlord has become aware of the defect.

Where the landlord becomes aware of defects and a tenant's breach of their repair obligations and OLA 1957 duty of care, then it is expected that the landlord will seek the enforcement of the tenant's obligations by serving the appropriate notices under the lease requiring the repairs to be affected. Where a tenant fails to affect the remedial works necessary to address the disrepair and alleviate the hazard; and where the landlord then has a right of entry into the property and a right to undertake "self-help" remedial works, then the landlord will be in breach of their duty of care if they fail to enter into the property and undertake remedial works on the tenants behalf as permitted by the lease.

Should an accident occur as a result of the landlord's want of action and breach of their duty of care, then any person injured or suffering a damage as a result may take action against the landlord (despite the tenant's liability under the lease. However, if the landlord suffers damages they will still have the right to seek damages recovery from the tenant under their contractual obligations. However, this may be easier said than done, particularly if the tenant is of weak covenant strength.

Dilapidations surveyors must understand the requirements of section 4 of the DPA 1972. Where dilapidations claims are made during the currency of a tenancy, the landlord should be made aware of the duty of care liability that they may take on under section 4 of the DPA 1972. The landlord should therefore consider the level of their public indemnity insurance according to the hazard and potential risk; and should seek to implement "self-help" works where the hazard poses a risk of a severe nature and of a reasonable likelihood.

Health and Safety at Work Act 1974

The Health and Safety at Work Act 1974 (HSWA 1974) has substantial implications for how "employers" occupy and use properties where the demised premises may form part of their workplace. Section 2 of the Act defines the general duty of employers to their employees.

The Act is also a parent/enabling Act for a broad range of subservient regulations, orders and byelaws that all relate to health and safety in the workplace. A great deal of health and safety-related subservient legislation will be of relevance to dilapidations and surveyors undertaking surveys in a property when they are considering the extent of tenants' compliance with their statutory obligations.

Surveyors acting on dilapidations should have a general understanding of:

- the Workplace (Health, Safety and Welfare) Regulations 1992
- the Regulatory Reform (Fire Safety) Order 2005
- the Health and Safety (Safety Signs and Signals) Regulations 1996
- the Gas Safety (Installation and Use) Regulations 1998
- the Electricity at Work Regulations 1989
- the Control of Asbestos Regulations 2006
- the Management of Health & Safety at Work Regulations 1999.

There are many other complex and extensive regulations and byelaws that may be of relevance during dilapidations inspections, but which may be outside the extent of the knowledge of the building surveyor. Where issues arising under the HSWA 1974 are suspected (particularly in relation to maintenance and safety of the property and building services), the surveyor should seek third party specialist advice from an appropriate health and safety consultant or other consultants, such as specialist mechanical and electrical consultants, so that statutory compliance and health and safety issues can be properly and adequately complied with.

Limitation Act 1980

The purpose behind this Act, which replaced an earlier "Statute of Limitations", was to prevent "stale" claims being brought to court after a certain period of time. The various sections of the Act detail the limitation timings for many different types of criminal and civil actions. The intention behind the Act is to set out a series of limiting "cut-off" periods where a party wishing to bring an action on a claim (such as under contract or in tort) has to bring the case to court within a set timeframe, otherwise they would lose the right to bring that claim, ie the claim becomes "time-barred".

In relation to dilapidations and service charge claims, these are either six or 12 years depending on how the lease contract was completed by the parties. The limitation period will be six years where a lease has been merely signed by the parties "under hand". More commonly, however, most commercial leases are a deed entered into "under seal"; and where this is the case, the limitation period for bringing a dilapidations claim before a court is for up to 12 years after lease end.

Part 3: Dilapidations common law principles of loss

The guiding principle on loss

In contract-based dilapidations claims, the Latin maxim of "restitutio in integrum" ("restoration to original condition") is the recognised and long-standing principle behind claims concerning breach of contract.

The courts have considered the principle of *restitutio in integrum* at length in many dilapidations cases. In the simplest of terms it has been held that the guiding principle on loss could be said to be:

> Where a party sustains a loss by reason of a breach of contract, he is, so far as money can do it, to be placed in the same situation with respect to damages, as if the contract had been performed ... [1]

Over the years, the courts have developed the guiding principle on loss and it is now well established that:

---

1 Parke B in *Robinson* v *Harman* [1843–60] All ER Rep 383 at 85.

> The quantum of damage is a question of fact, ...
>
> The fundamental basis is thus compensation for pecuniary loss naturally flowing from the breach; but ... (this) imposes on a plaintiff the duty of taking all reasonable steps to mitigate the loss consequent on the breach.[2]

Making a claim for dilapidations-related damages should also not be viewed as an opportunity for the landlord to "punish" a tenant. The landlord's objective is simple:

> you are not to enrich the party aggrieved; you are not to impoverish him; you are, so far as money can, to leave him in the same position as before.[3]

And that:

> (a Claimant) cannot be allowed to create a loss, which does not exist, in order to punish the defendants for their breach of contract. The basic rule of damages, to which exemplary damages are the only exception, is that they are compensatory not punitive.[4]

When making the claim, is should be remembered that:

> damages for breach of contract must reflect, as accurately as the circumstances allow, the loss which the claimant has sustained because he did not get what he bargained for ...[5]

The relevance of true intentions

The prospects of successfully gaining the damages claimed in any landlord's dilapidations claim will be greatly influenced by the landlord's true intentions for the subject property. The courts have considered the issue of intentions and have held that:

> In the case of ... damage to real property, this object [*restitutio in integrum*] is achieved by the application of ... [possible] different measures of

---

2 Viscount Haldane LC in *British Westinghouse Electric and Manufacturing Co Ltd* v *Underground Electric Railways Co of London Ltd* [1912] AC 673 at 688–68.

3 O'Connor LJ in *Murphy* v *Wexford County Council* [1921] 2 IR 230 at 240.

4 Megarry V-C in *Tito* v *Waddell (No 2)* [1977] Ch 106.

5 Lord Bridge of Harwich in *Ruxley Electronics and Construction Ltd* v *Forsyth* [1995] EGCS 117.

> damage. ... Which is appropriate will depend upon a number of factors, such as the plaintiff's future intentions as to the use of the property ...[6]

The courts have also sought to define the concept of intentions and have stated that:

> An "intention" ... connotes a state of affairs which the party intending ... does more than merely contemplate: it connotes a state of affairs which, on the contrary, he decides, so far as in him lies, to bring about, and which, in point of possibility, he has a reasonable prospect of being able to bring about, by his own action of volition. ...
>
> (and the intention must have) ... moved out of the zone of contemplation — out of the sphere of the tentative, the provisional and the exploratory — into the valley of decision ...
>
> ... Not merely is the term "intention" unsatisfied if the person professing it has too many hurdles to overcome, or too little control of events: it is equally inappropriate if at the material date that person is in effect not deciding to proceed but feeling his way and reserving his decision until he shall be in possession of financial data sufficient to enable him to determine whether the project will be commercially worthwhile.[7]

Also, as a direct result of many past dilapidations cases where claims have been made and pursued by landlords on the basis of false intentions, the courts have had cause to look at the genuineness of the landlord's intentions and have determined that:

> Intention, or the lack of it, to reinstate can have relevance ... to the extent of the loss which has been sustained.[8]

And that:

> the genuineness of the parties' indicated predilections can be a factor ... One of the factors that may be relevant is the genuineness of the plaintiff's desire to pursue the course which involves the higher cost.

---

6 Donaldson LJ in *Dodd Properties (Kent) Ltd* v *Canterbury City Council* [1980] 1 EGLR 15.

7 Asquith LJ in *Cunliffe* v *Goodman* [1950] 2 KB 237.

8 Lord Jauncey of Tullichettle in *Ruxley Electronics and Construction Ltd* v *Forsyth* [1995] EGCS 117.

> Absence of such desire (indicated by untruths about intention) may undermine the reasonableness of the higher cost measure.[9]

Ultimately, the courts have looked at false statements of intention and have found that:

> if the plaintiff has ... no intention of applying any damages towards carrying out the work contracted for, or its equivalent, ... It would be a mere pretence to say that this cost was a loss and so should be recoverable as damages.

When considering the claimant's intention, it is also important to remember that the intention being considered is that which existed at the material date of the claim (normally the date on which the lease was determined). However, if future intentions for the property change during the claim settlement negotiations, then any changed intention should be openly declared at the earliest opportunity.

The best way to prove that the landlord's claim based on the cost of intended remedial works is appropriate is for the landlord actually to commit to the works project and incur the costs. Failure to undertake the remedial works may therefore raise legitimate questions over actual intentions. That said though, even where the works have not yet been physically undertaken, the courts have accepted other arguments over sufficient "fixity of intention" to carry out the remedial works as claimed.

In judging claims, the courts aim to consider the fixity of landlords' intentions. For example, the landlord may have some strong personal interest in the subject matter; or may be under some duty to carry out the work; or may give an undertaking to carry out the work. Weakest of all, the landlord may simply claim that they intend to carry out the works. In each case, the landlord's evidence need not establish intention with certainty but it must carry conviction.

The reasonableness of the damages sought

It is an accepted common law principle in dilapidations claims that the cost of repairs is the "ordinary *prima facie*" measure of damages,

---

9 Lord Lloyd of Bewick in *Ruxley Electronics and Construction Ltd* v *Forsyth* [1995] EGCS 117.

subject to the caveat that there are no circumstances which would make it inapplicable or inappropriate as the measure of loss. However, even where the "ordinary measure" appears reasonable on initial appraisal, the courts have suggested that:

> The real question in each case is: What damage has the plaintiff really suffered from the breach?[10]

Consequentially, in addition to the issue of the landlord's genuine intentions the question of reasonableness of the landlord's intentions, action and claim must be considered. On this aspect, the courts have held that:

> the damages to be awarded are to be reasonable, that is as between the plaintiff [Claimant] on the one hand and the defendant on the other.[11]

And that:

> (it is a) principle that a [Claimant] cannot always insist on being placed in the same physical position as if the contract had been performed, where to do so would be unreasonable ...

In a leading modern House of Lords appeal judgment, on a claim concerning damages for breach of contract (and of relevance to dilapidations), the issue of reasonableness was further considered and it was observed that:

> the reasonableness of an award of damages is to be linked directly to the loss sustained. If it is unreasonable in a particular case to award the cost of reinstatement it must be because the loss sustained does not extend to the need to reinstate.

The courts continue to take the view that "the test of reasonableness has an important role to play"[12] in dilapidations claims.

---

10 Denning J in *Duke of Westminster* v *Swinton* [1948] 1 KB 524 at 534.
11 May J in *C R Taylor (Wholesale) Ltd* v *Hepworths Ltd* [1977] 2 EGLR 31.
12 Rix LJ in *Voaden* v *Champion* [2002] EWCA Civ 89.

The proportionality of costs to benefit

There is a further aspect to consider when assessing the reasonableness of a dilapidations claim, the issue of the proportionality of any works expenditure or costs being claimed as a loss. In some cases, the "ordinary" (cost of works) measure of the loss is not upheld because it could be said that:

> the cost of completion is grossly and unfairly out of proportion to the good to be attained. When that is true, the measure is the difference in value.[13]

The above principle originates from a judgment dating back to 1921 in the New York Court of Appeal and has since been much "celebrated" in dilapidations and contractual damages common law in England and Wales. In a leading modern House of Lords appeal judgment, on a claim concerning damages for breach of contract, having considered the 1921 case; it was held that it established "two principles":

> First, the cost of reinstatement is not the appropriate measure of damages if the expenditure would be out of all proportion to the benefit to be obtained, and, secondly, the appropriate measure of damages in such a case is the difference in value, even though it would result in a nominal award.[14]

Put simply, it would not be equitable or reasonable to spend, say, £60,000 in undertaking all the possible dilapidations remedial works at a property if a lesser selective package of targeted works of a value of, say, £45,000 could achieve the same end benefit; or if the damage to the reversionary interest would be only, say, £30,000 if the property was re-let in its current condition without undertaking the works; and where there is a reasonable prospect of being able to let the property on such terms.

In the above example, the landlord may well want to spend the £60,000, but it is not proportionate to the £30,000 diminution loss that they might otherwise suffer and so the landlord would have breached the duty to mitigate loss (see fn 2). If the landlord undertakes the more costly and unnecessary exercise then the landlord does so at risk and is

---

13 Cardozo J in *Jacob & Youngs* v *Kent* (1921) 129 NE 889.

14 Lord Lloyd of Bewick in *Ruxley Electronics and Construction Ltd* v *Forsyth* [1995] EGCS 117.

unlikely to recover the full expenditure. Such action could be considered to be an expensive act of vanity on the landlord's part, that goes against the principles of natural justice and any rash or reckless claim made by the landlord on this basis stands a good chance of failing.

Conclusion

In dilapidations claims, in addition to considering the influence of statute on a claim, any assessment of loss should have regard to the common law principles of how that loss should be calculated. This assessment of loss will need to consider not only the interpretation of the lease covenants or obligations and the merits of the possible related claims of breaches; but should also consider in more general terms:

- the guiding principle of *restitutio in integrum*
- the claimant's true intentions for the subject property
- the reasonableness of the claim
- the proportionality of the costs to benefit.

Before instinctively assuming that the cost of any works actually undertaken "must" be the measure of the loss, surveyors should take care to consider the common law principles of loss.

The current common law principles of loss can be surmised as favouring a position that the "ordinary" (cost of works) measure of the loss can only ever be the true measure of the loss where there is/was genuine intent to undertake the works (or they have been undertaken); and where it is/was reasonable to undertake the works; and where any expenditure on works is/was proportionate to any benefit to be gained. In all other circumstances such as where there is reason to question the intentions, reasonableness or proportionality of the landlord's actions and/or claim (or any combination of these factors), then the "true loss" should be measured by other means, such as measuring the diminution in value of the reversionary interest.

Part 4: The basis for dilapidation claims

Claims made under contract law

While the Statute of Frauds (1677) has been repealed, the legal principle for the transfer of an interest in land to be in writing remains and can now be found in section 2 of the Law of Property (Miscellaneous Provisions) Act 1989 (LP(MP)A 1989). A properly executed, signed and completed lease will ordinarily satisfy the requirements of section 2 of the LP(MP)A 1989 and will therefore constitute a legally binding contract between the lease parties. As a lease is by nature a contract between two parties, then it can be seen that any dilapidations claim made in relation to an alleged "breach of covenant" is in essence a claim for breach of contract and so the claim will be subject to contract law rights and remedies.

Where it can be shown that an actionable breach of contract (eg a breach of lease covenants and obligàtions) has occurred for which a tenant is liable, then a claim for contractual damages may be made in the normal manner, eg by identifying the lease clause breached; stating the cause (where known) and nature of the breach suffered; the appropriate remedial works sought to rectify the breach; and the value/cost of the remedy sought.

Where claims for actionable breaches are to be made, then the remedy available to the injured party (the party suffering the breach) will depend on whether the lease is continuing; or whether it has determined and the contract has ended.

If the remedy to be implemented involves a claim or action for damages, then the damages claim will be subject to the guiding principle of *restitutio in integrum* (eg financial "restoration to original condition"). The claimant, their surveyor and other advisers will therefore need to appraise and take into consideration various claim-specific damages restitution factors that will have a significant bearing on what "loss" can be legitimately claimed.

In terms of appraising any legitimate dilapidations damages, a claim will ordinarily be subject to statutory "caps" imposed by section 18(1) of the Landlord & Tenant Act 1927 (L&TA 1927).

In addition, any contractual damages claim will also be subject to the common law principles of loss (see Chapter 2, part 3 above).

Claims made under the law of waste

While most modern dilapidations claims are presented as contractual claims related to breaches of lease covenants, the ancient remedy of seeking damages for "waste" under the Statute of Marlborough 1267 remains available in many circumstances as an alternative (or complimentary) basis for a claim.

For example, where there is no contractual lease (such as when the tenant is a tenant at will or a tenant at sufferance), there will be no basis for a claim on the tenant under contract law and so the claim will need to be made for damages for waste. Even where there is a lease, it remains possible in some circumstances to make a claim under the law of waste, so long as the lease is "without special licence had by writing of covenant" that permits waste.

Such a "special licence" could be a landlord's clause containing an obligation on the landlord to repair. Alternatively, there may be lease clauses that expressly prohibit the tenant from causing waste, in which case claims for waste will be permissible.

There are four main types of "waste" worthy of consideration in the context of dilapidations claims.

- Voluntary waste: is the term given for waste brought about by a positive act that alters land or property to its detriment and causes damage. All tenants/occupiers are liable for voluntary waste.
- Permissive waste: is the term given for waste brought about by allowing land or property to deteriorate due to a want of attention (eg a failure to act or by an act of neglect). In modern times, only a tenant-for-years is now considered to be liable for permissive waste (see *Dayani* v *Bromley London Borough Council* [1999] 3 EGLR 144).
- Equitable waste: where a tenant for life or a tenant in tail commits wanton destruction to a property, they are said to be liable in equity for waste.
- Ameliorating waste: is the archaic term for an alteration to land or property that improves it. While not impossible, it would be unusual to find reasonable grounds to claim for dilapidations damages arising from an act that "improves" a property.

Where it can be shown that an actionable waste has occurred for which a tenant is liable, then a claim for damages under the law of waste may be made in the normal manner, eg by identifying the nature of the waste suffered; stating the appropriate remedial works sought to rectify the waste; and the value/cost of the remedy sought.

The occurrence of waste and, in particular, permissive waste is often difficult to prove. As a consequence, dilapidations claims made under the law of waste are now uncommon as there will normally be alternative and better remedies available to landlords under modern landlord and tenant law. It is always advisable to seek specialist legal opinion on such matters before making a claim for waste.

Dilapidations and Litigation

3

What is litigation?

A civil dispute is a dispute between two (or more) parties concerning a private matter, for example a dispute over a relationship; performance of a contract; a breach of a duty of care, etc. In dilapidations claims the parties to the civil dispute are landlord and tenant and the litigation relates to the breach of the lease covenant.

More often than not civil disputes will be resolved between the parties directly. Occasionally however, the parties find they are unable to resolve their differences and at that stage can consider other alternative means of resolution. If all other avenues of dispute resolution fail, then the parties' last resort is to commence litigation with an action in the courts.

A succinct definition of "litigation" is that put forward by J Maclean and J Scott in *The Dictionary of Building*, 4th ed (Penguin, 1995):

> Taking a matter under dispute before a court of law ... the evidence is given under oath and the decision is binding ...

Litigation in the context of dilapidations in England and Wales involves resolving disputes between one party and another through the courts, from commencement of proceedings up to the final trial. As a general rule, civil cases are heard before a judge sitting alone, whereas criminal cases are typically held before a jury.

When surveyors initiate the dilapidations claim they are at the very start of this legal process but all surveyors should be aware that each and every case could ultimately result in litigation and a court case in order to resolve the dispute.

Most dilapidations text books include a final chapter on litigation almost as an afterthought. Such an approach risks under-playing the significance of perhaps the most effective remedy available to a landlord or tenant when faced with a situation that appears irresolvable.

The court system in England and Wales

To understand litigation further, it helps to understand the nature of the court system in the UK.

A court of law and the associated legal system in England and Wales is different to that of Scotland and Ireland. Each has their own systems, legislation, common law and case law. The different legal systems in the UK are however all subject to the jurisdiction of the European Court of Justice, which requires that European legislation is incorporated into UK domestic law and legal system.

For the purposes of dilapidations and service charge disputes, this book considers the civil justice system in England and Wales only.

The courts operate a hierarchy system where courts operating at one level, such as the county courts, are required to follow judgments and legal principles laid down on similar cases at higher level courts such as the High Court, the Court of Appeal, the House of Lords, the Privy Council, and finally the European Court of Justice.

Civil cases commence in either the county court or High Court, largely depending on the value of the claim. Appeals (on points of law only) will be heard in the next higher court. Each judge has to follow the decision of a judge in a higher court, but is not bound by judgments in a court of equal or lower level. These higher level judgments that set down principles of justice are often referred to as "precedents" and collectively they form the body of what is called "common law".

What is the purpose of litigation?

There are a variety of remedies that the claimant can seek in the civil court system. A general understanding of the remedies available through the courts is essential for a building surveyor acting on dilapidations claims.

Should a civil dispute be litigated then the courts have discretionary powers to award or grant:

- the payment of damages
- the payment of a debt
- the payment of interest
- possession orders
- injunctions.

Damages

Damages for breach of contract arise when one party to a contract fails to perform an obligation set out in the contract. The purpose of the damages is forward looking in that it aims to place the wronged party in the position it would have been had the breach not occurred. An example of such a claim would be where a contractor failed to fulfil all obligations in a construction contract, eg a leaking roof due to sub-standard materials used, contrary to the architect's specification. The claim would be for a sum of money (damages) to correct the defect.

Damages for tort arise where injury, loss, or damage occurs to the claimant or the claimant's property. Contrary to contractual damages, damages in tort are backward looking in that the aim of the court is to put the wronged party in the position it would have been had the tort not occurred at all. An example of a claim in tort would be where a building surveyor negligently failed to spot subsidence in a building during a survey. The claim would be for damages to put the house purchaser in the position they would have been had they hired a building surveyor who was not negligent and who had spotted the subsidence, ie the costs of putting right the subsidence and associated costs and expenses.

Debt

Debt actions are closely linked to breach of contract claims, with the key difference being that debt actions are "liquidated", ie the exact sum is known and is specified in the documentation used to issue court proceedings. An example of a debt claim would be where a borrower defaulted on a bank loan or where a landlord, in enforcing a dilapidation remedy, pursued a *Jervis* v *Harris* [1996] 1 EGLR 78 self-help remedy in accordance with the lease and then sought the recovery of the cost of the works undertaken as a contractual debt.

Interest

Where the remedy sought by the claimant is damages or debt, the court can award interest on the sum claimed. This is a separate remedy because interest must be specifically claimed. The rules governing the actual prospects of recovery vary tremendously between contract and tort claims.

Possession orders

This is a non-monetary claim involving land or property. An example of such a claim would be where a property owner seeks possession from squatters. A specific and enforceable order is sought from the court by the claimant.

Injunctions

An injunction is either an order of the court requiring a party to perform a certain act, or an order restricting a party from performing a certain act. The remedy is discretionary in nature (ie the court can decide that the remedy is not appropriate) and the burden of proving that the remedy is suitable is on the claimant (ie the party seeking the injunction). An example of an application to the court for an injunction would be where a director in a construction business moved to another company and contacted their former clients in breach of a "restraint of trade" clause in the director's contract.

In the above examples there are potentially several roles for a building surveyor, ranging from the advisor dealing with the pre-litigation stage of a dispute, to the expert advisor to the litigant once proceedings are issued.

What are the rules of litigation?

The obligation on parties to comply with courts' procedural rules once litigation proceedings are commenced has always been substantial. Much court time is devoted to the interpretation of these procedural rules while considering the cases and the subsequent claims for costs.

Judges in civil litigation have two primary roles:

- first to make judgments on law

- second to make judgments on fact.

While the two areas are often blurred, it is the judgments on law that relate to the procedural rules, and a tactical awareness to interpret these rules to support one's case distinguishes an average litigator or surveyor from one that is exceptional.

It is therefore of paramount importance that surveyors acting on dilapidations claims have a good understanding of the rules and principles governing conduct and proceedings in civil disputes.

The background to the Civil Procedure Rules

The need for reform

Over the centuries, the courts have acted to interpret the will of parliament as set out in statute and apply the law. As both statute and common law has evolved so has the complexity and volume litigation in the courts. Over the years, the courts had become busier with lower grade litigation which the judges felt naturally should have settled at the pre-litigation stage.

Every so often, the courts find themselves weighed down by litigation or the judiciary and parliament recognise that the court system needs reform and modernisation and acts of parliament changing the civil litigation procedures are introduced.

Lord Woolf's review

The last major reform of civil procedure reform took place in the late 1990s to address perceived ills in civil dispute litigation. T Goriely, R Moorhead and P Abrams in *More Civil Justice? The Impact of the Woolf Reforms on Pre-action Behaviour* (The Law Society, 2002) observed that:

> Before 1999 parties could sleep walk through the procedural steps until just before trial, without focusing on the substantive issues.

In recognition of the civil litigation and court procedure problems that existed in the mid 1990s, Lord Woolf was then tasked with undertaking a root and branch review of the civil justice system in England and Wales.

Following substantial investigations and consultation with all those routinely involved in the litigation process, Lord Woolf published his report.

Lord Woolf said, among many conclusions drawn, that that there was too much litigation in England and Wales. The parties to a dispute had to be re-incentivised to avoid starting proceedings in the first place. He considered it essential that a dispute resolution culture should be encouraged that would seek to avoid litigation wherever possible.

Lord Woolf's fundamental objective was to open up the "black art" of litigation to allow settlement of disputes at the earliest stage possible. Lord Woolf also felt that the demarcation of duties prior to 1998 worked against openness, understanding, honesty, and thus early settlement. He therefore proposed reform of the entire process, from the commencement of a claim (eg from the initial claim letter before action) up to the final trial before a judge, and beyond, into the appeal procedure and applications costs.

Civil procedure reform

Lord Woolf's reform proposals lead to the introduction of the current Civil Procedure Rules (CPR). The Civil Procedures Act 1997 and the CPR 1998 (as amended 1999) were passed by parliament and were introduced in April 1999 to provide a procedural framework to all types of civil disputes, including dilapidations claims.

The CPR aimed to make dispute resolution more straightforward and less expensive than was often the case under the old county court rules and High Court rules, and reduce parties' scope to exploit the court system as part of case tactics, including for the purpose of delay.

The primary aim of the CPR was to resolve the problems identified by Lord Woolf in his review of the civil justice system; particularly the excessive costs, delays, and complexities which he felt were due to the entrenched adversarial nature of the litigation process from the very start of the claim. For example, in the case of a dilapidations dispute, the CPR applies from when the two parties first realise there is a dispute (and crucially this is a stage when commonly the building surveyors are the principal client advisor).

The CPR "overriding objective"

The CPR overriding objective is contained at Rule 1.1. It requires parties in a dispute to deal with the resolution of their differences in a way that saves expense, is expeditious and fair.

These primary principles are the very cornerstone of the CPR and the modern civil justice system. In addition to the overriding objective, many other rules were introduced which set out to guide parties, engaged in civil dispute, as to the court's expected standards of conduct and behaviour necessary to achieve the overriding objective. In particular, the CPR saw the formalisation and further development of good conduct and best procedural practices which were embodied in sets of "protocol" rules.

CPR "protocols"

Within Part 1 of the CPR, there is a requirement that parties to a civil dispute follow a reasonable procedure intended to avoid litigation (a "pre-action protocol").

Pre-action protocols are applicable to all types of dispute at the pre-litigation stage, and aimed at reducing the number of court proceedings issued. Each protocol is a significant means by which the CPR Part 1 "overriding objective" is achieved and sets a broadly applicable programme for the resolution of civil disputes. The protocol sets out a series of procedural steps that a potential claimant in litigation must address before they can litigate.

The purpose of the protocol is to provide simultaneously a "brake" to litigation that is not ready for the courts; and secondly to ensure that if a dispute is litigated then the extent of the disclosure process undertaken pre-lease end will be such that the dispute may be settled with all the key facts known to the parties. A party does not have to abide by the protocol if they decide not to; but they may be penalised if they ignore it, even in cases where they are seen as the "winning party". To achieve this end, the CPR protocol provided a procedural timetable of actions that parties involved in a dispute had to comply with before they commenced court proceedings.

Originally, in 1998, there were just two subject-specific protocols (for injury and medical negligence claims) focusing on areas of civil dispute perceived as problematic by the courts due to the high volume of litigation generated. Between 1998 and 2006, eight other subject-specific protocols were approved by the Department for Constitutional

Affairs (now the Ministry of Justice) relating to key areas of litigation that the courts considered required their own protocol due to the high volumes of litigation in various sectors. However, commercial property dilapidation disputes, which do not yet have a formally "approved" protocol, are just one of the many areas of civil dispute covered by the more general CPR protocol.

The CPR, and in particular the protocols, are highly relevant to surveyors who are very often the principal professional advisors during the initial negotiations following the service of the schedule of dilapidation, ie the document that sets out a detailed claim for damages based on an allegation of breach of contract.

When a surveyor follows the approach set out in the CPR default protocol, adherence to the basic principles will make the claim fairer, quicker and more economic. This will have benefits in terms of efficiency of dispute resolution for both surveyors and lawyers, and most importantly direct cost savings for their landlord and tenant clients.

The success of the CPR

The legal framework created by the CPR binds all parties (and their advisors including surveyors and valuers, etc) during the pre and post litigation stages of a dispute.

The CPR revolutionised the way that disputes of a civil nature are handled. Patterns of behaviour of litigants, lawyers and surveyors, together with their general conduct during proceedings from the outset of a claim are materially relevant to the courts; and the cost consequences of improper conduct should not be lightly dismissed.

The reforms have in the main achieved Lord Woolf's desired objective of reducing the number of cases issued out of the courts. The Royal Institution of Chartered Surveyors (RICS) Dispute Resolution Faculty (2002) neatly summarise the aims of the CPR as:

> modernising the courts, speeding up procedures, and reducing the cost of litigation.

Dilapidations protocols

Surveyors should appreciate that the CPR extended the obligations to all persons involved in all civil disputes (including surveyors), requiring them to comply with procedural rules at all stages before litigation.

The incidental consequences of the CPR reforms in dilapidations-related civil disputes have effectively given building surveyors a "quasi litigator" role, ground previously held solely by lawyers. Surveyors are therefore an unusual hybrid of expert and advocate, with obligations that must be carefully addressed.

Issues concerning the appropriate procedure for dilapidations must be clarified in order for surveyors to better understand the place of dilapidations within the CPR framework.

The PLA's unadopted "dilapidations protocol"

Prior to the CPR 1998, the minimum professional standard of conduct and service expected of surveyors acting on dilapidations claims was dictated predominantly by the RICS *Dilapidations Guidance Note.*

In 2000, the Property Litigation Association (PLA) published a proposed "protocol" specifically covering dilapidations claims. It sought to build upon the established RICS guidance and to further develop best practice from a litigation perspective. Once published, the PLA protocol was put forward for court approval in 2002. It was generally supported by lawyers as a positive step, and also by RICS who subsequently included a copy of the early draft PLA protocol within the appendices of the fourth edition of the RICS *Dilapidations Guidance Note.*

The original PLA protocol failed to gain formal court "approval" and, following a further review and revision by the PLA, a revised dilapidations protocol was published by the PLA in September 2006, again with the view that the courts should formally approve it. While many aspects of the 2006 version of the protocol were welcomed, significant parts of the proposed protocol were less well received by surveyors. In May 2008, a further revised version of the PLA protocol was released but this is considered by some as still having scope for further improvement.

The, as yet, unapproved status of the PLA dilapidations protocol (version 3) does not, however, mean that there is no applicable or binding protocol when attending to dilapidations claims. Surveyors must bear in mind that there is an alternative officially approved protocol in the guise of the CPR "default" pre-action protocol.

The CPR default pre-action protocol

Within Part 1 of the CPR, there is a requirement that parties to a civil dispute follow a reasonable procedure intended to avoid litigation. Notably, dilapidations and service charge disputes are not litigation topics that have thus far warranted the introduction of formally approved specific protocols.

With the aim of promoting the overriding objective of the CPR in areas of civil litigation currently without a court "approved" protocol, the CPR created a default pre-action protocol. This default protocol is set out in section 4 of the CPR "Practice Directions" on protocols.

This was confirmed by the then Lord Chancellor's Department (now Ministry of Justice) entitled Civil Justice Reform Evaluation (August 2002) stating that:

> The Practice Direction on Protocols requires parties to comply with the general spirit of the protocols whatever the subject of the claim.

The Practice Direction on Protocols published by the Ministry of Justice (including revision 43 as of March 2008) requires adherence with the following default pre-action protocol rules when acting on dilapidations cases. These default pre-action rules relate to many types of disputes, including dilapidations, and are repeated below:

> 4.1 In cases not covered by any approved protocol, the court will expect the parties, in accordance with the overriding objective and the matters referred to in CPR 1.1(2)(a), (b) and (c), to act reasonably in exchanging information and documents relevant to the claim and generally in trying to avoid the necessity for the start of proceedings.
>
> 4.2 Parties to a potential dispute should follow a reasonable procedure, suitable to their particular circumstances, which is intended to avoid litigation. The procedure should not be regarded as a prelude to inevitable litigation. It should normally include —
>
> (a) the claimant writing to give details of the claim;
> (b) the defendant acknowledging the claim letter promptly;
> (c) the defendant giving within a reasonable time a detailed written response; and
> (d) The parties conducting genuine and reasonable negotiations with a view to settling the claim economically and without court proceedings.

4.3 The claimant's letter should —

(a) give sufficient concise details to enable the recipient to understand and investigate the claim without extensive further information;
(b) enclose copies of the essential documents which the claimant relies on;
(c) ask for a prompt acknowledgement of the letter, followed by a full written response within a reasonable stated period; (For many claims, a normal reasonable period for a full response may be one month.)
(d) state whether court proceedings will be issued if the full response is not received within the stated period;
(e) identify and ask for copies of any essential documents, not in his possession, which the claimant wishes to see;
(f) state (if this is so) that the claimant wishes to enter into mediation or another alternative method of dispute resolution; and
(g) draw attention to the court's powers to impose sanctions for failure to comply with this practice direction and, if the recipient is likely to be unrepresented, enclose a copy of this practice direction.

4.4 The defendant should acknowledge the claimant's letter in writing within 21 days of receiving it. The acknowledgement should state when the defendant will give a full written response. If the time for this is longer than the period stated by the claimant, the defendant should give reasons why a longer period is needed.

4.5 The defendant's full written response should as appropriate —

(a) accept the claim in whole or in part and make proposals for settlement; or
(b) state that the claim is not accepted.
(c) If the claim is accepted in part only, the response should make clear which part is accepted and which part is not accepted.

4.6 If the defendant does not accept the claim or part of it, the response should —

(a) give detailed reasons why the claim is not accepted, identifying which of the claimant's contentions are accepted and which are in dispute;
(b) enclose copies of the essential documents which the defendant relies on;

(c) enclose copies of documents asked for by the claimant, or explain why they are not enclosed;
(d) identify and ask for copies of any further essential documents, not in his possession, which the defendant wishes to see; and
(e) (The claimant should provide these within a reasonably short time or explain in writing why he is not doing so.)
(f) state whether the defendant is prepared to enter into mediation or another alternative method of dispute resolution.

4.7 The parties should consider whether some form of alternative dispute resolution procedure would be more suitable than litigation, and if so, endeavour to agree which form to adopt. Both the Claimant and Defendant may be required by the Court to provide evidence that alternative means of resolving their dispute were considered. The Courts take the view that litigation should be a last resort, and that claims should not be issued prematurely when a settlement is still actively being explored. Parties are warned that if this paragraph is not followed then the court must have regard to such conduct when determining costs;

It is not practicable in this Practice Direction to address in detail how the parties might decide which method to adopt to resolve their particular dispute. However, summarised below are some of the options for resolving disputes without litigation:

- Discussion and negotiation.
- Early neutral evaluation by an independent third party (for example, a lawyer experienced in that field or an individual experienced in the subject matter of the claim).
- Mediation — a form of facilitated negotiation assisted by an independent neutral party.

The Legal Services Commission has published a booklet on 'Alternatives to Court', CLS Direct Information Leaflet 23 (*www.clsdirect.org.uk/legalhelp/leaflet23.jsp*), which lists a number of organisations that provide alternative dispute resolution services.

It is expressly recognised that no party can or should be forced to mediate or enter into any form of ADR.

4.8 Documents disclosed by either party in accordance with this practice direction may not be used for any purpose other than resolving the dispute, unless the other party agrees.

4.9 The resolution of some claims, but by no means all, may need help from an expert. If an expert is needed, the parties should wherever possible and to save expense engage an agreed expert.

4.10 Parties should be aware that, if the matter proceeds to litigation, the court may not allow the use of an expert's report, and that the cost of it is not always recoverable.

These rules are directly applicable to dilapidations claims to be prepared and served in England and Wales. In contrast to the PLA's unadopted Dilapidations Protocol (2008), the CPR default protocol as set out above is legally binding.

The CPR default protocol is non-prescriptive in terms of timescales for service of a landlord claim and makes useful reference to the appointment of "experts", even acknowledging that joint experts are not going to be useful in all cases, which can often be the case in a dilapidations claim.

If the CPR default protocol is read in conjunction with the RICS *Dilapidations Guidance Note*, then a surveyor will have the pre-litigation issues relating to dilapidations dispute resolution predominantly covered. All surveyors acting on dilapidations cases should pay due regard to this CPR default protocol.

In May 2008, RICS submitted a response to the Ministry of Justice consultations on protocols and confirmed that the default protocol is appropriate for use in dilapidations claims. This was a landmark decision by RICS and opened up the debate regarding the future of the specialism.

Civil or criminal breaches

When considering the civil dispute framework, practitioners should be aware that in certain circumstances, the disrepair of a property or way in which a claim is conducted may be considered to be a breach of a statute that results in a criminally punishable offence for their clients, in addition but separate to any original contractual breach and claim under civil law. Such instances may occur when the premises include, or are exclusively, residential and is covered by statute intended to protect the health and wellbeing of the residents of those premises.

Whenever practitioners are involved with the possibility of a prosecution for disrepair of residential premises (or premises that include an element of residential), then they should take prompt action as the practitioner and their client may face a criminal prosecution for such disrepair.

The legislation governing the instances where a landlord may be in breach of the criminal law, include section 369(5), as amended by the Local Government and Housing Act 1989. Any breach of these statutory requirements may lead to a fine at the local magistrates' court. The case may be brought by the local authority against the company which has failed to keep the premises in repair, as well as a named person, such as the company secretary or a director. Section 613(1) of the Housing Act 1985 permits named persons to also be charged where they are personally seen as liable for the disrepair.

It is important to note that the charges brought against either or both of the parties (company itself or a named person) list each of the items said to be in breach (similar to separate items of alleged breaches on a Scott schedule) and a separate charge is brought for each item. Therefore, if there are, say, 60 alleged breaches, which would be listed on a Scott schedule, then 60 charges are brought against the parties said to be liable. Each charge has a maximum penalty of £5,000.

The Role of Professional Advisors

4

The importance of professional advice

The nature of dilapidations claims is complex, with specialist professional advice crucial to achieve the correct settlement. On all but the simplest or smallest of disputes, the landlord and tenant should consider obtaining advice from suitably experienced professionals. Disputes where the "dilapidations team" functions as a cohesive and communicative unit will tend to produce fairer and quicker settlements, which is the ultimate goal of most parties involved in such a dispute, particularly the paying landlord or tenant client.

All professional advisors acting on a case are expected to provide their client with the benefits of their specialist knowledge and expertise. Advisors should also seek to protect their client from unethical behaviour or overly partisan conduct.

Professional advisor: "impartial expert" or hired gun?

Pre-CPR 1998

There is a widespread myth that pre 26 April 1999 (when the Civil Procedure Rules (CPR) came into force), consultants acting as expert witnesses were allowed to act as the "hired gun", to be partisan in their reporting and to favour their client. Even pre-CPR it was a breach of Rules of the Supreme Court 1965 and the County Court Rules 1981 for an expert to be partial.

The reality of pre-CPR was that surveyors and other consultant/ contractor advisors appointed for the benefit of their expertise should have been impartial in providing their services and advice. Even pre-CPR, they should have being ever mindful that they could end up giving evidence or testimony in a court and that they, therefore, had an overriding duty to the court.

Cresswell J set out what is regarded as a classic distillation of these principles in *National Justice Compania Naviera SA* v *Prudential Assurance Co Ltd (The Ikarian Reefer)* [1993] 2 EGLR 183:

> The duties and responsibilities of expert witnesses in civil cases include the following:
>
> 1. Expert evidence presented to the court should be, and should be seen to be, the independent product of the expert uninfluenced as to form or content by the exigencies of litigation. (See *Whitehouse* v *Jordan* [1981] 1 WLR 246)
> 2. An expert witness should provide independent assistance to the court by way of objective, unbiased opinion in relation to matters within his expertise. An expert witness in the High Court should never assume the role of an advocate.

While the CPR now has better defined pre-litigation procedures and overriding objectives, there has not been a fundamental change in the expert advisor's role since introduction of the CPR in 1999.

Post-CPR 1998

Both landlord and tenant clients alike are still often amazed, disappointed and frustrated at the levels of apparent exaggeration applied to dilapidations claims by surveyors and other consultants in relation to a dispute that was quite simply a claim for civil damages.

Surveyors and other advisors who work on the basis that the CPR only applies when a claim is litigated and therefore feel able to gamble with pre-litigation negotiation positions, must re-read the rules. The simple fact is that they have no choice on the point at which the CPR applies. It is applicable from the very inception of a dispute.

The perceived reason why surveyors (and other advisors generally to a lesser degree) don't behave in accordance with the CPR and therefore do not take it seriously is because the vast majority of claims settle out-of-court (although often at questionable levels). Their advisor's behaviour and often unreasonable conduct, therefore, never gets subjected to scrutiny or criticism in open court. Furthermore, this poor professional approach rarely results in disciplinary referrals of the surveyor or consultant to their professional body and so the erroneous or questionable practices are allowed to perpetuate.

The pre-litigation "expert" advisor

In *Abbey National Mortgages plc* v *Key Surveyors Nationwide Ltd* [1995] 2 EGLR 134, Judge Hicks QC defined an expert as:

> expert, in relation to any question arising in a cause or matter, means any person who has such knowledge or experience of, or in connection with that question, that his opinion on it would be admissible in evidence.

Professional advisors therefore are arguably "experts" from the start of their involvement in a dispute because of the following.

1. In almost every case, they are preparing a document in contemplation of litigation. If the claim does not settle then the landlord will be using the schedule for the purposes of court proceedings and the surveyor will probably be required to be an expert witness.
2. The CPR Default Pre-action Protocol states at Part 4.9:

> The resolution of some claims, but by no means all, may need help from an expert. If an expert is needed, the parties should wherever possible and to save expense engage an agreed expert.

In the context of dilapidations claims, the surveyor appointed by the landlord or tenant is arguably the "expert" envisaged by the CPR Default Protocol, ie the independent person who will offer expertise on an impartial basis in order to resolve the claim

There are perhaps a few exceptions to the rule that surveyors and advisors involved in dilapidations must always be appointed as independent experts. A dilapidations surveyor could be an advisor where there is no cause of action, ie where the default position is not litigation. Examples of such situations include certain instances of:

- lease assignments
- lease surrender negotiations
- conditional lease break negotiations.

However, where litigation is a possibility, then regardless of the surveyor's professional title or the label that the RICS attach to the perceived role when providing their services (be it surveyor, negotiator, advocate or expert), the surveyor will be acting in an "expert" capacity from the outset.

The CPR "expert" in court proceedings

The precise role of the surveyor or valuer appointed to be an "expert" in a dispute will be crystallised at the point at which the landlord client decides to litigate, or when the tenant client cannot reach settlement and the opponent is forced to issue proceedings.

The role is governed by the Civil Procedure Rules 1998 which formalise the requirement of impartiality and reasonableness in respect of anyone holding themselves out as having the requisite expertise to aid the court in the resolution of a dispute. There is a potential role as an "expert" in dilapidations claims for any professional person with the appropriate level of experience in relation to the dispute in question, including:

- building surveyor
- valuer
- quantity surveyor
- engineers
- contractor.

A key principle of the role of an expert is that two experts can hold differing views, both of which can be reasonably held. The difference between these positions is referred to by lawyers as the "band of reasonableness" and explains why parties involved in a dispute can both be represented by "experts" and yet still be in disagreement.

Part 35 of the CPR (sections 35.1–35.3) set out below, best summarises the requirements of an "expert":

> 35.1: Expert evidence shall be restricted to that which is reasonably required to resolve the proceedings.
>
> 35.2 A reference to an 'expert' in this Part is a reference to an expert who has been instructed to give or prepare evidence for the purpose of court proceedings.

35.3(i) It is the duty of an expert to help the court on the matters within his expertise.

35.3(ii) This duty overrides any obligation to the person from whom he has received instructions or by whom he is paid.

From the outset, surveyors and lawyers dealing with dilapidations claims should carefully factor in the requirements of CPR Part 35 and the associated Practice Direction, which states:

1.1: It is the duty of an expert to help the court on matters within his own expertise: rule 35.3(1). This duty is paramount and overrides any obligation to the person from whom the expert has received instructions or by whom he is paid: rule 35.3(2).

1.2: Expert evidence should be the independent product of the expert uninfluenced by the pressures of litigation.

1.3: An expert should assist the court by providing objective, unbiased opinion on matters within his expertise, and should not assume the role of an advocate.

1.4: An expert should consider all material facts, including those which might detract from his opinion.

1.5: An expert should make it clear:

(a) when a question or issue falls outside his expertise; and
(b) when he is not able to reach a definite opinion, for example because he has insufficient information.

1.6: If, after producing a report, an expert changes his view on any material matter, such change of view should be communicated to all the parties without delay, and when appropriate to the court.

Although the initial schedule at lease end is technically not an expert witness report, it is a document prepared by an expert in contemplation of civil proceedings. The survey report (ie the schedule of dilapidations or counter schedule) is prepared by the surveyor appointed by the landlord/tenant on the basis of their expertise. If the claim does not settle, the document will become the basis of the expert report for litigation and any unexplained change of position will reduce the credibility of the documentation produced by that individual.

If the surveyor acting for the landlord or tenant pays close attention to Part 35 and the Practice Direction, this will give a solid base to any claim from the outset.

Impartial advisor

If a claim proceeds to litigation then the expert's opinion expressed in the pre-litigation stage will be open to review and cross examination. Under the CPR, written reports are prepared following discussions between experts which open up further risks for the building surveyor, particularly when the report cannot be materially altered by the instructing solicitor.

The role and professional stance adopted by the surveyor has a crucial impact on the way that a claim is handled and therefore on the ability of the respective advisors to reach a reasonable and timely settlement. The ambiguous nature of the role of the professional advisor during the pre-litigation stage needs to be addressed and carefully managed on each claim.

It should not be forgotten that two comparable advisors (surveyors, engineers, valuers, etc) can be impartial and independent experts, yet both hold different professional views and opinions on the same subject.

Crucially, by defining the role as independent experts, the division between the parties will tend to narrow.

Surveyors need to be encouraged away from extreme positions by the formalisation of their appointment as an "expert". By focusing the mind of the surveyor on the contents of the schedule or counter schedule, clarity can be achieved and weak arguments exposed.

Asking the advisors to disclose the terms of their appointment and to sign a statement of truth on their opinion of the claim or defence, will often flush out exaggerated positions. Providing that the duty is equally applied to the other party to the dispute, the duty imposed will tend towards the settlement of the dispute.

Ultimately though, the advisors remain just that and ideally the landlord or tenant party to the claim should themselves also be asked to sign and warrant their stated intentions within the claim documents.

The building surveyor

Building surveyors generally become involved at the pre-litigation stage when acting for clients during the early stages of a dispute, often when the parties have not yet openly disagreed. An example of this is a landlord client instructing a building surveyor to prepare a terminal schedule of dilapidations.

Timing, knowledge and tactics are crucial, because while most building surveyors will tread carefully when they are a court-ordered "expert witness" in a contentious case that will be litigated, many building surveyors are less aware of their professional responsibilities before that point is reached.

The role of the building surveyor

The current fifth edition of the Royal Institution of Chartered Surveyors (RICS) *Dilapidations Guidance Note* is ambiguous on the exact role a surveyor should play in terms of their appointment, positioning surveyors in the uncomfortable middle ground between "expert" and "advisor". These are very different roles (one supposedly impartial and the other by definition partisan) and can easily become unworkable and contradictory; are open to abuse; and will prejudice the claim if litigated.

The RICS Guidance Note explains that a surveyor acting as an advisor should be objective and impartial when producing a document that must be suitable for litigation and that the surveyor should be "influenced by the considerations relating to expert witnesses". Quite what this flexible phraseology requires in practice is not clear and such fence-sitting serves only to obscure the quite simple "expert advisor" role of the surveyor.

Building surveyors dealing with dilapidations often become involved in a professional capacity before the lawyer is instructed to offer procedural advice. They may be the first point of contact for a concerned client, or their investigative work may initially highlight the problem.

Building surveyors are subsequently required to have a good general level of understanding of:

- building pathology and construction technology
- the nature and characteristics of landlord and tenant interests in property
- the basic principles of claims under contract law and the law of waste
- the relevant statute and common law relating to dilapidations
- the civil litigation framework
- the rights and remedies available to the parties
- the CPR and the Default Protocol.

Building surveyors and the CPR

The CPR imposed obligations on building surveyors to comply with procedural rules from the outset, which are particularly relevant to building surveyors in two ways.

1. Building surveyors represent their clients at the "pre-litigation" stage and must understand that their conduct early on can have a lasting impact even at final trial.

2. Building surveyors are often instructed as "expert witnesses" in litigation. The Court of Appeal in *Field* v *Leeds City Council* [2000] 1 EGLR 54 set out the very straightforward requirements for an expert:

   > the "expert" must: (1) demonstrate that they have relevant expertise in an area in issue in the case; and (2) demonstrate that he/she is aware of their primary duty to the court if they give expert evidence.

Post-CPR, the determination of liability in a dispute is rarely left until the final trial. Where the experts were once cross-examined by the respective party's barrister, post-CPR, most disputes are settled earlier between the experts.

As the system for civil litigation has moved towards the surveyor being a central player and sometimes quasi arbitrator in the litigation process, an increased level of responsibility rests on the expert's shoulders. However, with responsibility comes risk.

Building surveyors should be aware of the consequences of the CPR both when acting as an expert in civil litigation cases and when involved in their day-to-day work where litigation is not contemplated. Where a dispute exists and an expert's report is required, experts are required to sign a "statement of truth" in relation to any evidence they give. With a risk of imprisonment for contempt of court, perjury, or a criminal act of fraud under the Fraud Act 2006, the surveyor ought reasonably to have known that the contents of the statement they were making was either reckless or untrue.

Merely reminding an opposition expert that the CPR has teeth can simultaneously unnerve an opponent, impress a client, and gain an advantage. Building surveyors are common suppliers of such evidence, eg in dilapidation claims, rights of light dispute, and construction disputes.

With adequate knowledge, building surveyors can both avoid problems for themselves and obtain a tactical and therefore commercial advantage for their client.

The building surveyor's professional duty

A building surveyor appointed to represent a landlord or tenant client, should work on the basis that they are an "independent expert" from the outset of a claim. All correspondence and any communication via the Scott schedule format should be issued on the basis that it will eventually be read by the trial judge. This may not end up being the case, but it is a suitable starting point.

The proverbial stick to encourage such an approach is that this will serve to protect the surveyor and the client in the event that the matter is litigated. The surveyor who adopts an "expert-like" approach from the outset, avoiding the one-sided interpretation of all issues at every opportunity, will have a schedule or response document and associated chain of correspondence that stands them and their client in good stead. This will leave them ready to be formally appointed as the CPR Part 35 expert when proceedings are contemplated and then commenced.

Such an approach ensures that the surveyor will avoid the wrath of the client and the judge if the schedule or response documents are one-sided to the extent that they are considered to have exacerbated the dispute and incurred costs.

The proverbial "carrot" in relation to this approach is that a moderate and well-reasoned position, if adopted by both sides, will get to the root of the dispute earlier, and if maintained by both sides, will resolve the dispute quicker in most circumstances. This will be considered positively by the client and judge in the unlikely event that the matter is not resolved prior to litigation.

Professional standards of service

All RICS chartered building surveyors acting on dilapidations claims are expected to comply with the current RICS *Dilapidations Guidance Note* (currently fifth edition as of June 2008). The raison d'etre of any RICS guidance note is covered by the following RICS statement found in the forward of each note:

> RICS Guidance Notes provide advice to members of the RICS on aspects of the profession and procedures that embody best practice and meet a high standard of professional competence. When an allegation of professional negligence is made against a surveyor, the court is likely to take account of the contents of any relevant RICS Guidance Notes as they are relevant to professional competence.

The RICS *Dilapidations Guidance Note* is not an all-encompassing text book on every aspect of a dilapidations claim, just a useful reminder to practitioners about how a dilapidations dispute should be handled in the broadest of terms; and a benchmark for professional negligence in the worst cases.

The building surveyor's fees and terms of appointment

Because a dilapidations claim is a legal damages claim (therefore by definition governed by the CPR from the outset), all schedules should be prepared on the basis that they are done so "in contemplation of litigation", ie as experts. Consequentially, any incentive-based or performance-related fees are incompatible with the role of the surveyor dealing with dilapidations claims.

Landlord and tenant clients do not want their potential expert witnesses to have to change their fee agreements just prior to litigation, at the risk of distorting the perception of independence of all documentation produced up to that date.

The lawyer

The tactical preparation and execution of a dilapidations claim or defence is normally greatly enhanced by the involvement of a lawyer at an early stage. For example, the lawyer may be able to provide beneficial advice with regards to tactical timing, available claim remedies or defences, litigation costs protection measures and interpretation of covenants.

In particular, lawyers and building surveyors should seek to work together from the outset of a claim. Between them, they should advise their client to their claim or defence strategy before formal claim or defence documents are served, in what could otherwise be considered a rash manner.

A lawyer's fee for service of the landlord's schedule on the tenant, are often relatively minimal and generally recoverable under the lease, subject to the inclusion of an appropriate lease clause at the start of the lease. Furthermore, using a lawyer ensures that the document is served on the correct address at the correct time and involves the lawyer at this initial stage of the damages claim. Surveyors should be encouraged to involve a lawyer for this purpose as a matter of best practice.

Sadly, a claimant or defendant often sees the early involvement of a lawyer as an additional and unnecessary cost and so often the first involvement of a lawyer in any dilapidations issue is when they serve or acknowledge a claim prepared by surveyors.

During the pre-litigation dispute

At whatever stage they become involved, the lawyer should take the primary legal advisory role and should review and seek amendments, changes, addition or clarification of any material, but draft claim or defence documents prepared by other advisors prior to their service.

The negotiation stage, which is often initially handled by the surveyor, progresses more effectively when the lawyer is involved to advise upon matters requiring legal interpretation; or where the benefit of their expertise may allow the identification of a legal argument that either strengthens or weakens the case so that subsequent attendances and cost risk can be mitigated.

The lawyer should seek to play "devil's advocate" within the advisors team, to test the strength of any claim, defence argument or rational. Their involvement may also become necessary or be essential to ease the dispute through points of impasse that arise between appointed surveyors (where perhaps common sense and a rational approach to dealing has become the first victim of professional intransigence, improper conduct or ego-driven opinions).

The lawyer may also need to take a leading role in the preparation and timing of pre-litigation measures, such as proposing or considering alternative dispute resolution (ADR) solutions; or the pre-litigation preparation and service of any of the CPR Part 36 without prejudice settlement proposals and other strategies to mitigate cost risks.

During litigation

If matters proceed to litigation then the lawyer will become the key advisor and should assume the management and co-ordination role within the team of professional advisors.

At litigation stage, it will be essential for the lawyer to advise on further expert legal appointments and to ensure that the court processes and protocols are followed so far as is possible on any given case. They should also seek to co-ordinate and oversee any "expert" appointments.

The lawyers will need to ensure that court orders are complied with; and that court-ordered reports are submitted in time. They will normally also co-ordinate the review of privilege correspondence and the timely disclosure of documents. Where difficulties arise in complying with court order timings, the lawyers will be expected to deal with procedural matters and applications for extensions of time, etc.

Up until the date of the hearing in court, the lawyer may also need to take a leading role in the preparation and timing of pre-litigation measures, such as proposing or considering ADR solutions; or the preparation and service of any of the CPR Part 36 without prejudice settlement proposals and other strategies to mitigate cost risks.

At settlement

Once a settlement is achieved in principle between the parties, the default position should again be that lawyers are instructed to formalise the settlement in legally appropriate terms. Most settlements warrant a lawyer's involvement to ensure that all liabilities are addressed in a way that concludes the relevant liabilities as intended by the parties.

Even a small claim can become complicated and expensive if the legal agreement is ambiguous. Depending on the value of the claim, the lawyer may also need to undertake Financial Services Authority (FSA) money laundering checks before accepting a settlement on behalf of a client.

In long-running and complex cases, the issues of litigation costs may also be extensive and difficult to resolve in their own right and may need a hearing on costs to resolve. In such circumstances, the lawyer will be expected to co-ordinate and prepare all costs claims.

The claim settlement issues are beyond the remit and areas of expertise of most surveyors, and so the default position should again always be that a surveyor should recommend a lawyer's involvement to conclude the dispute.

The valuer

Not all dilapidations claims require formal valuation evidence. This basic premise is confirmed in the Property Litigation Association (PLA) *Dilapidations Protocol* (2008) and the current RICS *Dilapidations Guidance Note*. However, valuation issues must always be considered in relation to dilapidations claims, to ensure that the landlord is claiming only for what they have lost (even if the consideration given merely confirms that the commissioning of a formal diminution valuation is not felt necessary in the circumstances).

Section 18(1) of the Landlord and Tenant Act 1927 "first limb" diminution of the landlord's reversionary interest valuations, are merely a statutory confirmation of the common law principles of loss, ie that a claimant can only recover from the defendant what they have actually lost. Both the diminution valuation and the common law loss assessment are of vital importance to any claim. One will inevitably limit the other.

The role of the valuer is often separate to that of the building surveyor, on the basis that if done correctly it requires a different commercial valuation expertise more commonly held by a chartered general practice surveyor.

Ideally, the valuer should be familiar with the local market and the property type and experienced in the principles and realities of dilapidations. Because valuation evidence is normally formally commissioned after the preparation of the schedule, it is important that the technical aspects of the claim are clearly communicated to the valuer and that they are provided with a full copy of any dilapidations claim and/or defence in existence at the date of their valuation.

It is preferable that a meeting should be arranged on- site between the building surveyor and the valuer to agree their approach, as this will help the valuer understand the claim and should help avoid discrepancies or contradictions in the way that the claim is handled.

At the point of instruction, the valuer should be issued with the following documents.

1. Lease and licence documents.
2. Schedules and counter schedules.
3. Relevant evidence for the actual or potential letting of the demise in question, including:
   (a) market appraisals

   (b) heads of terms and lease agreements for the demise in question (or others in the building if multi-let, or nearby).
4. Related correspondence between the landlord and tenant surveyors.
5. Valuation reports issued by the opposing party to the dispute.
6. Copies of any written advice from lawyers and counsel.

It should be remembered that any valuation evidence or reports gained will be materially relevant and disclosable documents and so copies of any valuation report obtained should be made available in a timely manner once received.

Also, as valuation is seen as an "art" rather than an exact science, the commissioning of a valuation by one side to the dispute will often lead to the other side obtaining their own valuation. Therefore, a "battle" of the valuers often follows, where the courts require an opinion of a third wholly independent valuer. Consequentially, if contemplating commissioning a valuation report, it is best for both parties to agree to commission a single valuation from a single joint expert valuer in keeping with the CPR Default Protocol. This will have the benefit of keeping valuation costs and subsequent dispute to a minimum.

Other professional advisors

During their survey of a property, the surveyor will ordinarily identify the extent and necessity of the attendances and input into the claim or defence of other consultant advisors or contractors.

For example, in larger commercial claims, mechanical and electrical plant and other building services-related issues often equate to 25–50% of a claim initially collated by the surveyor. Many dilapidated buildings also have structural issues that need specialist investigation and advice. Where the above circumstances arise, it may be necessary to seek the advice of a mechanical and electrical engineer or structural engineer.

Few building surveyors and lawyers are qualified or sufficiently competent to address all the potential specialist issues arising at lease end in a commercial building, and so third party consultant input or specialist contractor investigation and testing is often required.

Terms of appointment of any third party consultant or contractor used by a surveyor in a dilapidations claim should be set out clearly and be CPR compliant. The sub-consultant should work on the same impartial "expert" basis and this must be documented.

5 What is the Leasehold Interest?

An introduction to tenancies

Introduction

An understanding of applicable dilapidations remedies is related to the degree of exposure that a surveyor has to the variety of types of tenancy.

The relevance of the type of tenancy

Dilapidations claims are predominantly prepared and negotiated by building surveyors, often without the full involvement of the landlord's solicitors or agents. Because most building surveyors are not familiar with all the various types of tenancies that exist, they may not be aware of the appropriate dilapidations remedies available for the given type of tenancy. This in turn can lead to building surveyors commencing inappropriate dilapidations claims to the detriment of their client.

Establishing the tenancy type

Given the relevance of the type of tenancy in a dilapidations claim, a surveyor should never take the "tenant" information at face value and should seek to undertake some basic due diligence checks (see Chapter 9).

Initial due diligence checks should not only look at the "tenant" under the lease but the interest/identity of the landlord and any sub-tenants or other building occupants. The types of interest that may be found can be generally grouped into three classes:

- freehold interests
- leasehold interests
- informal tenancies.

In keeping with the various tenancy definitions as generally summarised by HM Revenue & Customs (HMRC), set out in the section below, are the most common types of tenancy types that a surveyor may come across during landlord and tenant proceedings.

Freehold interests

Tenant in fee simple

A freeholder is a tenant in fee simple. It is the nearest anyone can get to absolute ownership of land.

Joint tenants

Joint tenants have equal rights to possess the same property. They acquired it from the same source. When a joint tenant dies, the survivor takes their share. When only one person survives, they become entitled absolutely to the property.

Tenants in common

Tenants in common are effectively co-owners of an undivided beneficial interest in land. This gives them possession of it. No individual tenant in common has exclusive possession of any part of the property. When a tenant in common dies, their successor becomes a tenant in common in their place. The interest does not pass to the other co-owners.

Tenancy in common usually refers to land. It can also refer to moveable property.

Leasehold interests

Tenant for life

A tenant for life has a life interest in an estate. The term particularly applies to an interest in settled land.

The legal estate in the land is vested in the tenant. This is as trustee for all the beneficiaries. (This does not make the tenant a trustee of the settlement). However, the tenant is also a beneficiary. This entitles the tenant to the benefits from the land.

The tenant is restricted by the doctrine of waste. This imposes restrictions on how the tenant for life can use the land.

The tenant has the power to sell, lease, or deal with the land as a full owner. The tenant must exercise these rights and powers in good faith. The tenant has to act for the benefit of others.

Tenant in tail

A tenant in tail has a life interest in an entailed estate (settle-landed estates on a specific line of heirs).

Tenant for years

A tenancy for years or a "fixed-term tenancy" is a tenancy or lease for a fixed period or term.

The date of commencement and the length of a lease must be agreed before there can be a legally binding lease. The term can also be for a fixed period of less than a whole year.

It may take effect from the date of the grant, an earlier date, or a date up to 21 years ahead.

At the end of the fixed-term, the lease or tenancy comes to an end automatically: there is no need for a notice to quit.

However, if the tenancy is a residential "assured tenancy", it will continue at the end of the term as a statutory periodic tenancy, unless it is brought to an end by surrender of tenancy or a court order.

Tenant from year-to-year

A tenancy from year-to-year (or month-to-month or less) is, as the name may suggest, a short-term tenancy and is one that ordinarily will automatically renew until appropriate notice and action is taken by one of the tenancy parties to terminate the interest. Such a tenancy will ordinarily have comparatively lesser tenant obligation than is the case for the other types of leasehold interests.

Informal tenancies

Tenant at sufferance

A tenancy at sufferance arises when an occupant or former tenant remains within a property without a lease or a right to "hold over" and the landlord has not indicated whether or not they agree to the tenant's continued occupation. If the landlord gives their express agreement and recognises the occupant as a tenant, the occupant becomes a tenant at will.

Tenant at will

A tenancy at will is a tenancy that can be terminated by the landlord or the tenant at any time.

A tenancy at will usually arises by implication, when the owner of land allows a person to occupy it although they have no fixed-term, periodic tenancy, or licence (for example, when a landlord agrees to the tenant holding over).

More rarely, a tenancy at will may be created by express agreement.

A tenancy at will of business premises does not have the statutory protection given to a business tenancy. In the case of residential premises, however, the usual statutory protection from eviction will apply. A tenancy at will can be terminated by the landlord demanding possession or if either the landlord or the tenant dies or parts with their interest in the land.

Tenant by estoppel

A tenancy by estoppel is a tenancy that exists despite the fact that the person who granted it had no legal right to do so. Such a tenancy is binding on the landlord and tenant but not on anyone else. If the landlord subsequently acquires the right to grant the tenancy, it automatically becomes a full legal tenancy.

Other tenancy terms and definitions

Business tenancy

A business tenancy is a tenancy of premises that are occupied for the purposes of a trade, profession, or employment.

Business tenants have special statutory protection. If the landlord serves a notice to quit, the tenant can usually apply to the courts for a new tenancy. If the landlord wishes to oppose the grant of a new tenancy, they must show that they have statutory grounds, which may include breaches of the tenant's obligations under the tenancy agreement or the provision of suitable alternative accommodation by the landlord. Otherwise, the court will grant a new tenancy on whatever terms the parties agree or, if they cannot agree, on whatever terms the court considers reasonable. When the tenancy ends, the tenant may claim compensation for any improvements they have made.

Under the Landlord and Tenant (Covenants) Act 1995, in force from 1 January 1996, when business tenancies are assigned the new tenant generally takes over the covenants (or promises and warranties) of the first tenant in the lease, except when otherwise agreed. Previously the old tenant was always liable, even after assignment, if a subsequent tenant defaulted on the lease.

Long leases

A long lease is a lease that lasts more than 50 years. This is the distinction between a lease that is defined as an occupational lease of premises. Clearly, a long lease can contain both dilapidations obligations and service charge provisions, but it is not seen as an occupational lease applying the guidance of HMRC, which see these lease lengths as not being for occupational purposes under the taxation rules it has in place.

A lease of a residential flat would normally be longer than 50 years (in order to qualify for a mortgage when purchasing the flat) and this also would contain service charge provisions, eg cleaning and security of the common parts of the premises.

Short leases

Short leases are leases for periods of less than seven years. Short leases can be granted by private landlords, registered social landlords (such as housing associations) or public bodies such as local councils.

They include periodic tenancies where the tenant has not got a fixed-term agreement and occupies property on, eg a weekly or monthly basis. In some cases the arrangement between the parties will not be a tenancy, but a licence.

Periodic tenancy

If a landlord starts to accept "rent" on a regular basis from a tenant at will, then an ordinary periodic tenancy is created. A tenant at will with a periodic tenancy gains the statutory "security of tenure" protection afforded to tenants under the Landlord and Tenant Act 1954.

Assured tenancies

An assured tenancy is a form of tenancy under the Housing Act 1988 that is at a market rent but gives security of tenure. The premises may be furnished or unfurnished. This kind of tenancy replaces protected tenancies, except those in existence before the Housing Act 1988 came into force. Former assured tenancies under the Housing Act 1980 (where different provisions applied) are converted into the new kind of assured tenancy.

To qualify as an assured tenancy, the premises must be let as a separate dwelling, within certain rateable value limits. There are certain exceptions, such as when the landlord lives in another part of the same premises.

When the tenant of an assured tenancy dies, the tenant's spouse has a right, in circumstances, to take over the tenancy as successor to the deceased tenant. An assured tenant cannot usually assign the tenancy without the landlord's consent.

Assured shorthold tenancies

An assured shorthold tenancy (AST) is a special kind of assured tenancy, at the end of which the landlord is entitled to recover possession without having to show one of the usual grounds for possession of an assured tenancy. This kind of tenancy was introduced by the Housing Act 1988, replacing protected shorthold tenancies.

Under the Housing Act 1996, from 28 February 1997 all new residential tenancies are assured shorthold tenancies without security of tenure, unless a notice is specifically served stating that the parties are creating an assured tenancy.

6 Dilapidations Remedies

An overview of remedies

No matter how onerous a tenant's obligations under a lease are, the landlord will be afforded little protection if they lack the ability to take effective remedial action in the event of tenant breaches.

The remedy options available to a landlord for addressing any physical and/or financial damage, for which the tenant is considered liable, will depend on a number of factors such as:

- the nature of the tenancy
- whether the tenancy is ongoing or has determined
- the nature, extent and severity of any dilapidations and/or breaches found at the property
- the nature and amicability of the landlord and tenant relationship.

Where there is a formal lease with agreed terms, covenants and obligations lease or tenancy has yet to determine, the landlord's available remedies may include one or more of the following options:

- To seek forfeiture of the lease and financial damages via the courts.
- To peaceably "re-enter" into the demised property, thereby determining of the lease and seeking financial damages.
- To allow the lease to continue but to seek financial damages for the breaches in lieu of compliance remedial works.
- To seek a court order for "specific performance".

- To seek voluntary compliance with the lease by the tenant, with the option for the landlord to undertake "self-help" remedial works enforcement action in the event of continued tenant default.
- To seek a "surrender" of the lease (recission of the contract) in return for a suitable compensation payment to be made for the lost rent opportunity and any dilapidations damages.
- To do nothing and to defer enforcement action or damages claims to a later stage in the lease.

Where there is no formal lease, then the landlord's available remedies may include one or more of the following options:

- Where applicable, to seek to formalise the lease arrangements and agree terms in accordance with the procedures of the Landlord and Tenant Act 1954; and to include within any new lease an obligation on the tenant for the undertaking of the current dilapidations remedial works at the commencement of the new lease term.
- To seek to regain the premises and to claim damages for waste to the extent permitted by the law of waste.

Where the lease or tenancy has irrevocably determined, the only remedy available to the landlord will be to seek a financial payment of damages as compensation for losses directly attributable to breaches of the lease and/or under the law of waste.

Remedies available during the lease

Forfeiture

A frequently encountered remedy is that of the landlord serving a notice on the tenant under section 146 of the Law of Property Act (LPA) 1925 seeking forfeiture of the lease as a consequence of the tenant's breaches. The most common occasion where you find forfeiture being sought, is in relation to breaches of tenant's covenants for commercial premises; or in relation to "long lease" residential tenancies.

The concept of Forfeiture

The concept of Forfeiture is based on the doctrine of "re-entry", where the landlord "forfeits" or terminates the tenancy by re-entering into or onto the property. As soon as the landlord re-enters the property, the tenancy terminates.

The statute and common law in forfeiture controls how and when a landlord may take action to terminate a tenancy as a direct consequence of a tenant being in breach of a covenant or condition of their lease.

Statutory limitations on forfeiture

The statute and common laws apply to forfeiture proceedings in relation to commercial, residential and agricultural tenancies in general, although subject to slightly different legal rules and procedures depending on the nature and type of the tenancy.

There are various procedures and statutory restrictions on executing forfeiture proceedings, such as those contained in section 146(1) of the LPA 1925; or section 31 of the Rent Act 1965; or sections 167–171 of the Commonhold and Leasehold Reform Act 2002.

Once the landlord has commenced and/properly executed a claim for forfeiture, they will ordinarily cease to treat the tenant as a "tenant". Should the landlord continue to do so, the doctrine of "waiver" may apply and the tenant may strengthen their grounds to challenge the forfeiture action.

It is important to understand that when the landlord commences proceedings for forfeiture in court, a "constructive" re-entry takes place on the date the proceedings are served (and not on the date when a court makes an order to such an effect).

Assuming that the forfeiture process has been properly implemented prior to the court hearing, the nature of the proceedings are not to "terminate" the tenancy (as forfeiture has by then already occurred) but for the landlord to recover possession of the premises.

Where a landlord successfully gains a court award for forfeiture, the tenant will normally be held liable for financial damages to compensate for the losses the landlord suffers as a result of the early determination of the lease. The quantum of damages may include damages for the loss of future rental income and further compensation for the dilapidations-related losses (where the dilapidations element remains subject to the statutory and common law principles of loss for dilapidations).

Tenant relief from forfeiture under the Leasehold Property (Repairs) Act 1938

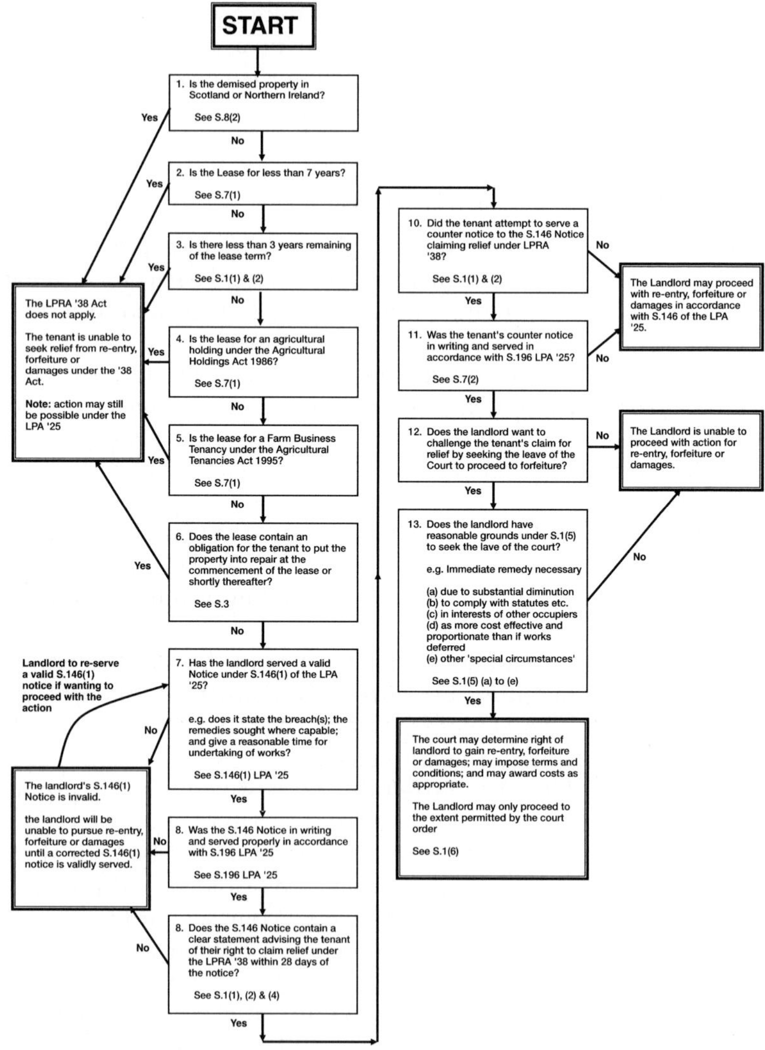

Forfeiture of commercial tenancies

With regards to forfeiture of commercial tenancies, the procedures for commencing forfeiture proceedings as set out in section 146 of the LPA 1925 apply.

Section 146(1) of the LPA 1925 requires the landlord to serve notice on the tenant:

- specifying the particular breach complained of
- if the breach is capable of remedy, requiring the lessee to remedy the breach
- in any case, requiring the lessee to make compensation in money for the breach.

Under section 146(1) of the LPA 1925 and section 18(2) of the Landlord and Tenant Act 1927 (L&TA 1927), the landlord is also obliged to allow the tenant a reasonable period in which they are to comply with the notice and to undertake the remedial works.

Commercial tenant's "relief" from forfeiture

Due to historic abuse of the forfeiture rights and remedies by landlords, statute and common law has evolved over the years to provide a fair and balanced degree of protection to reasonable tenants who may only be in breach of their tenancy obligations in a relatively inconsequential way.

In general, if the landlord fails to seek forfeiture in strict compliance with statutory processes and restrictions; and/or if the tenant considers that the termination of the tenancy would be unduly harsh in the context of the severity of any breach, then a tenant in receipt of section 146(1) of the LPA 1925 may seek relief from forfeiture from the courts under section 146(2) of the LPA 1925.

Also, in some circumstances, the tenant may able to claim the "benefit" of the Leasehold Property (Repairs) Act 1938 (LP(R)A 1938) and gain "relief" without having to apply to the courts for relief. However, the LP(R)A 1938 is often misunderstood and the legislation has a number of conditions and restrictions that may make the benefit of the Act unavailable to a tenant (see the flowchart opposite).

When faced with a claim for forfeiture, the onus is upon the tenant to bring about a claim for relief if they do not wish the tenancy to be forfeited. Relief will generally be available to the tenant during forfeiture proceedings.

Forfeiture action can also normally be defended (and the tenancy would be reinstated) once the tenant has made a payment to the landlord of all outstanding sums due under the lease; and/or commits to remedying of outstanding breaches of tenant's obligations or covenants.

Forfeiture of residential tenancies

If the property is a residential property, then alternative rules will apply for forfeiture such as those set out in sections 167–171 of the Commonhold and Leasehold Reform Act 2002.

However, the focus of this book is to consider commercial dilapidations issues and a surveyor acting for landlords contemplating forfeiture action on a residential tenant should seek further legal advice from the landlord's solicitor before commencing such an action.

The "re-entry" remedy

The landlord's right of re-entry

The remedy of re-entry is normally available where the lease includes express landlord's rights, allowing them to peaceably "re-enter" into the premises in the event of a tenant's breach of the lease; and to regain possession thereby determining the lease.

The process to be followed by the landlord and the circumstances in which this option is available to the landlord will normally be described within the lease itself. However, as with forfeiture action, the process remains subject to statutory and common law procedures and rules.

For example, a landlord wishing to determine the lease by implementing the re-entry procedures in a commercial lease will be required to serve advance written notice upon the tenant in accordance with section 146(1) of the LPA 1925.

A tenant's right of relief

A commercial tenant in receipt of a section 146(1) notice may claim relief from re-entry under section 146(2) of the LPA 1925 or under the LP(R)A 1938 (as per the rights of relief for against forfeiture — see above).

Re-entry termination of the lease

For commercial leases, following the expiry of a valid and properly served section 146(1) notice, if the tenant fails to claim relief in the required manner (either under section 146(2) of the LPA 1925; or where applicable under the LP(R)A 1938) then the landlord may execute the re-entry and peaceably re-enter into the premises without the leave of the court. In doing so the lease will immediately determine and the commercial tenant will become liable for damages in the same manner described above for forfeiture action.

Again, there are separate rules and statutory procedures that apply to residential tenancies and where the landlord is contemplating forfeiture action on a residential tenant, the surveyor should seek further legal advice from the landlord's solicitor before commencing such an action.

Orders for "specific performance"

An introduction to specific performance

The remedy of specific performance is the term given to the remedy process of making an application to the courts to gain a court order directing the landlord and tenant to specifically and materially perform the covenants and obligations of a lease.

Specific performance is considered to be an "equitable remedy" and is not available as a right will only be granted at the court's discretion.

When asked to issue an order for specific performance, the courts will seek to exercise their discretionary powers having first had due regard to long established principles. The key principles a court will consider are:

- adequacy of damages as an alternative remedy
- the "mutuality of remedy"
- the extent and degree of court supervision required
- whether the specific performance is being sought for only a part; or all of the breached covenants and obligations that could require performance.

The adequacy of damages

When considering an application for specific performance, the court will look at the adequacy of damages as an alternative remedy. It has been held that a court will only give specific performance in preference to damages when specific performance "can...do more perfect and complete justice" — see *Wilson* v *Northampton and Banbury Junction Railway Co* (1873–74) LR 9 Ch App 279.

The mutuality of remedy

When considering the request for specific performance, the courts will also take into consideration the mutuality of remedy as defined by Buckley LJ in *Price* v *Strange* [1977] 2 EGLR 49 where it was said that:

> it should have become an accepted rule that equity would not compel one party to perform these obligations specifically in accordance with the terms of the contract unless it could also ensure that any unperformed obligations of the other party would also be performed specifically.

In other words, it would be unfair to expect one party to fully and specifically perform their obligations while the other party remains at liberty to disregard their obligations.

Court supervision

The court will also seek to take into consideration the extent of court supervision that would be associated with any order for specific performance; and will ordinarily be disinclined to grant specific performance where the court would be required to be constantly in attendance.

The need for full performance

Finally, when considering an application for specific performance, the court will seek to ascertain whether or not the application concerns specific performance of only part or the whole of the contractual obligations.

The courts will not normally grant specific performance for only part of a contract (although in certain limited circumstances a mandatory injunction may be gained for part performance of a contract).

Extended application

Until 1998, the remedy of specific performance had only been available to the tenant seeking enforcement of the landlord's obligations upon the landlord.

However, in the case of *Rainbow Estates Ltd* v *Tokenhold Ltd* [1998] 2 EGLR 34, the courts ordered specific performance in the landlord's favour for the first time due to unusual case-specific circumstances. In doing so it set a precedent that the remedy is available to landlords as well as tenants, albeit in very limited and rare circumstances.

Ordinarily, landlords will have alternative and as effective remedies available to them under the terms of a lease, and so it remains highly unlikely that a landlord will be able to gain an order for specific performance. If the landlord in a dilapidations case wishes to seek specific performance, then they should be encouraged to take specialist solicitor or counsel opinion before commencing such an action.

"Self-help" enforcement action

The concept of "self-help" action

In most modern commercial and residential long leases, the lease will contain obligations on the tenants to undertake various acts of repairs, decoration and maintenance, etc; either at prescribed dates or periods in the lease; or as an ongoing perpetual obligation.

In most modern leases, there will also be express clauses permitting the landlord to inspect the premises and to thereafter serve a notice upon the tenant identifying apparent breaches of the repair, maintenance and/or decoration obligations, etc. The lease clauses will normally require the tenant in receipt of such a notice to commence and/or undertake the works necessary to remedy the breaches within a pre-defined period. This type of notice is commonly called a "notice to repair".

Where the tenant receives a notice to repair and then fails (in part or wholly) to comply with the notice; then most leases will grant consequential rights for the landlord to "enter" into or onto the premises (with surveyors and contractors, etc); and to undertake the necessary works to remedy the tenant's remaining breaches that were originally identified within the notice. A landlord in doing so will then ordinarily be permitted by the lease to seek immediate repayment from the tenant of costs incurred as a commercial debt on an indemnity basis.

In this type of remedy and action, because the landlord takes positive action to help protect their property interests, it is commonly referred to as "self-help" action. Alternatively, the same action may be referred to as "*Jervis* v *Harris*" action after the Court of Appeal judgment which upheld and confirmed a landlord's self-help rights in the case of *Jervis* v *Harris* [1996] 1 EGLR 78.

Distinction between entry and re-entry

It is important to understand that self-help action and the exercising of a right of "entry" is subtly different to the lease terminating action of "re-entry".

The self-help right of entry is only a temporary action by the landlord that lasts only as long as it takes for the landlord to undertake the remedial works in the tenant's default. The landlord's subsequent action in seeking to regain costs incurred is as a commercial debt and is not for damages. If the self-help remedy is fully implemented, the lease continues unaffected for the remainder of the term. Whereas by comparison, the alternative remedy of re-entry results in the lease determining and an action for compensatory damages.

Limited tenant's rights of relief

Furthermore, because the landlord does not intend to determine the lease by way of forfeiture or re-entry; and is also not seeking "damages", a commercial tenant will be unable to seek "relief" under section 146(2) of the LPA 1925 and is also unable to claim the benefit and relief under the LP(R)A 1938.

However, under section 147(1) of the LPA 1925, the tenant may apply to the county courts for relief from "self-help" notice to repair works relating to wants of internal decorative repair if they satisfy the conditions of the Act and make a court application for relief prior to the landlord commencing such works. Unfortunately, the cost involved in seeking relief from the county court in many moderate cases may outweigh the cost of undertaking the decorations and so, in practice, relief under section 147(1) is rarely sought.

"Unpriced" self-help schedules of dilapidations

It is worth noting that there is no requirement to serve a costed or "priced" schedule of dilapidations with the landlord's notice to repair.

The intention of the remedial action is that the remedial works will be undertaken. As either the tenant's or landlord's tendering contractors will quote for the works, any landlord's surveyor's estimated costing or pricing of the schedule of dilapidations would be entirely academic. Furthermore, any attendance and costs charged by the landlord's surveyor in producing an academic and hypothetical set of prices for the schedule works would be unreasonable and disproportionate; and would be subject to challenge by a well-advised tenant.

Risks of landlord's "self-help" remedy

If a landlord is considering implementing a dilapidation self-help remedy, the surveyor should inform the landlord of the risks inherent with the action.

For example, the extent of "self-help" works that the landlord may be entitled to undertake will normally be determined by the express self-help clause within the lease. This clause may not necessarily be drafted as wide as the other lease clauses concerning repair, maintenance and decoration, etc. As a result, if the self-help clause only grants a landlord the right to undertake "repairs", then the landlord can't extend this right to permit the undertaking of, say, wants of reinstatement of unauthorised alterations where such breaches exist.

Should a landlord attempt to exceed the permitted self-help works, then they are at risk of breaching the tenant's quiet enjoyment and may find themselves subject to court action and/or an injunction halting their transgressing works.

The landlord will also have to incur the works costs (and professional fees) in the first instance before they may be demanded from the tenant. The tenant may lack the funds to settle the debt promptly (or at all) and the landlord may be left out of pocket for a considerable period or even permanently.

On most occasions, the tenant will also seek to delay or disrupt the landlord's self-help works as they will normally wish to avoid the debts that will be incurred, particularly if it is disproportionate to any loss that would be suffered by the landlord if the works were not undertaken.

Self-help advantages for the landlord

The main benefit to the landlord is that if the landlord enters into the property and properly and reasonably executes self-help works following the tenant's non-observance or performance of the notice to repair, the landlord can seek the full costs incurred from the tenant as a debt.

The commercial debt action is not currently subject to any statutory or common law restrictions relating to dilapidations damages actions, such as the restrictions imposed by section 18(1) of the L&TA 1927.

To further illustrate the advantages of self-help action for a landlord, it should be considered in comparison to the benefit to be had by a landlord in undertaking an alternative remedy such as re-entry and damages.

For the purpose of illustration, let's say that the gross costs of the remedial works necessary for a tenant to fully comply with the lease repair, maintenance and decoration, etc obligations was, say, £100,000. Let's also suppose that the damages associated with the tenant's breaches only result in a £30,000 diminution to the landlord's reversionary interests.

Under self-help action, the landlord could undertake all of the remedial works if the tenant failed to do so and recover the full £100,000 worth of costs as a debt. Whereas under the damages route, it would be considered unreasonable to expend £100,000 on remedying the lesser £30,000 damage suffered and the landlord would have to make up the £70,000 difference at their own expense.

Future self-help restrictions?

It is speculated that it is probably only a matter of time before the arguably inequitable cost-to-damage situation that may arise from self-help action are challenged in a court higher than the Court of Appeal; and if challenged, may well be overturned. In the meantime, whilst it may appear inequitable for the tenant, self-help action is currently perfectly legitimate to pursue and implement and surveyors should not be afraid to implement an effective means of protecting their client's property interests.

Surrender deals

As a properly completed lease is, by its nature, a contract, an often overlooked remedy for tenant breach and non-compliance is to seek a "rescission" of contract.

Rescission means the cancellation of a contract by mutual agreement of the parties. When dealing with lease contracts, this is more commonly referred to as agreeing a "surrender" of the lease. Normally, a surrender of the lease will only be worth considering and proposed by the landlord where the market conditions are in the landlord's favour.

If there is only a single tenant under the lease without sub-tenants, then the surrender negotiations may seek to surrender the lease without granting a new lease; or there may be a surrender and partial re-letting proposed (for example a retail tenant may wish to surrender the upper residential parts to a property, but retain the ground floor commercial parts).

Even if there are sub-tenancies to any lease, then the surrender of the lease option remains available without the surrender or determination of any sub-leases subject to the requirements of section 150 of the LPA 1925.

Favourable circumstances for surrenders

It may be advantageous to consider a surrender deal where the tenant is experiencing financial difficulties and where self-help debt actions or damages action would be likely to have a financial impact that the tenant could not afford. If the tenant is in a dire financial position, it may be better for the landlord to agree a pragmatic surrender deal or longer term programme of planned remedial works and maintenance rather than risk the tenant becoming bankrupt or financially collapsing, resulting in a potentially unrecoverable loss.

Circumstances when it may be in the landlord's favour to consider a tenant's proposed surrender, could be when there are reasonable prospects of finding a new tenant for the property who is of similar or better covenant strength and who is willing to pay at least the same level of rent; and/or where they would be willing to commit to similar or more favourable lease terms). It may also be advantageous to agree a surrender deal where the landlord could put the property to a better or alternative use; or where the landlord could redevelop the site and/or property in order to gain an increased capital value.

Surrender premium considerations

Where a surrender deal is a favourable option, the terms of the deal will normally include a surrender premium by the party gaining the most benefit from the early termination of the lease. If the tenant is to pay a surrender premium, then this will normally consist of a premium in compensation for loss of rent and rates, etc and a premium sum as compensation towards dilapidations. Alternatively, if the landlord is seeking the surrender for their benefit, they may need to pay compensation to the tenant and may also have to forego the ability to claim for dilapidations.

In agreeing surrender in return for a compensation premium, there may also be tax or VAT advantages to both parties in how the various surrender premium elements are structured. The landlord and tenant parties may therefore be able to agree a premium structure that is of greater benefit than pursuing a confrontation claim and seeking damages.

Surrender negotiation risks

Where the lease parties are willing to consider the surrender of a lease, care should be taken to ensure that alternative dilapidations and lease rights and remedies are not "waived" under section 148 of the LPA 1925 or become otherwise legally prejudiced.

All correspondence between the parties and their surveyors and legal advisers on issues of surrender should be issued on a "without prejudice" and "subject to agreement" basis. The parties should also openly reserve their other dilapidations rights and remedies in case the surrender deal collapses.

Damages during the tenancy

The right to accept damages

There may be occasions where the landlord will be willing to accept a payment from the tenant of financial damages during the term of the lease or tenancy, in lieu of the tenant undertaking breach or waste-related remedial works; and without the landlord seeking forfeiture or re-entry.

Also, where it can be shown that an actionable waste has occurred for which a tenant is liable, then a claim for damages may be made under the law of waste.

The difficulty in seeking damages

In principle, the settling of a dilapidations claim during a lease or tenancy term will be an option, although in practice it will be difficult to quantify the level of damages (if any) due.

In most circumstances, tenants with commercial leases will be required to pay rent at market rates that are normally appraised and set during rent reviews on the assumption that the property is in full repair and that there is no diminution to the reversionary value related to dilapidations. Consequently, for damages claims made during the lease term, under section 18(1) of the L&TA 1927, any landlord seeking dilapidations-related damages during the lease term will struggle to be able to demonstrate that they have suffered any damage to their reversionary interest or that any damages are due.

Alternatively, if the damages claim is being made under the Law of Waste, while the claim will not be subject to any statutory restriction imposed by section 18(1) of the L&TA 1927, the claim for damages will still be subject to the common law principles of loss under the Law of Waste. Again, it may be difficult for the landlord to demonstrate that they have suffered any damage to the reversionary interest or that any damages are due.

The consequences of accepting damages

If a landlord can demonstrate that there has been a financial damage suffered during the term as a result of dilapidations, then there will be future dilapidations claims consequences in accepting damages during the term. Once the landlord accepts damages during the term for breaches that exist at the date of the damages payment, then the landlord has achieved restitution for the breach and the tenant will no longer be liable for the ongoing breach for the remainder of the lease.

If there is further deterioration and dilapidations, then the landlord may still make a claim for further damages at a later stage, but the courts will ordinarily deduct any damages already paid by the tenant and so the recoverable damages in the later claim will be limited to the damages caused by the further deterioration only.

These further damages may be near impossible to quantify or substantiate because once the landlord accepts damages in lieu of works, if they do not then expend those damages gained in undertaking the original dilapidations, then it is highly probable that further consequential waste and deterioration will occur as a result of the landlord's permissive actions which the tenant will not be liable for.

Generally, it is found that the disadvantages in landlords accepting damages in lieu of remedial works during the term greatly outweighs any advantage and so very few landlords seek damages during the term without also seeking to determine the lease by forfeiture or re-entry action.

If the landlord on a dilapidations claim wishes to seek damages in lieu of works during the lease term and without seeking to determine the lease through forfeiture or re-entry, then the surveyor should advise the landlord to take specialist legal opinion before proceeding.

Injunctions for waste

Where there is no lease (such as where the tenant is a tenant at will or tenant at sufferance) and where actionable waste is being suffered during the currency of the tenancy, then the landlord's remedy during is to seek an injunction against the tenant, restraining the tenant from either continuing the waste or from commissioning further waste.

If the tenant's waste is serious enough to warrant seeking an injunction then, normally, the landlord will also seek to terminate the tenancy and recover the property (then seek damages). Alternatively, the landlord may seek to formalise the tenancy by requiring the tenant to commit to a lease (which will then make the more normal contractual dilapidations remedies available to the landlord to pursue).

Lease end damages

Damages being the only remedy

Once a lease or tenancy determines (however so determined), the landlord and tenant interest will come to an end and the contractual relations and obligations imposed by the lease will also cease at the date of determination.

After the lease or tenancy has determined, the only dilapidations remedy available to the landlord will be to seek a payment of damages as compensation for any loss or damages suffered attributable to a breach of the lease, or for waste occasioned to the property during the term of the lease or tenancy.

The lease end dilapidations claims made under contract law will be subject to the restrictions imposed by section 18(1) of the L&TA 1927 and the wider common law principles of loss (see Chapter 2).

Conduct of the surveyor

Dilapidations claims to be made after the end of the lease or tenancy are expected to be made within a reasonable period after the determination of the lease or tenancy.

Currently, it is accepted good practice to seek to make the claims within, say, eight weeks of the end of the term, although case-specific circumstances may not always allow this. There is also a statutory limitation on bringing claims for contractual damages for breaches of the lease of a maximum of 12 years from the end of the lease under the Limitation Act 1980 (or six years where the lease is merely signed under hand rather than being executed as a deed).

All end of term dilapidations damages claims will be claims made in contemplation of litigation, and so the claims should be conducted so far as possible in accordance with the Civil Procedures Rules (CPR) and the CPR Practice Directions (see Chapter 3).

7 Surveyor Due Diligence

Part 1: Appointing the surveyor

Many of the difficulties that arise at later stages in dilapidations proceedings stem directly from a poor understanding of the surveyor's professional duty or inadequate consideration at the surveyor's initial appointment.

Prevention is better than a cure. Therefore, in an effort to seek to minimise the potential for conflict, the surveyor should consider initial appointment issues such as their:

- professional and statutory duty of care
- ethical duty of care
- ability to provide independent services
- possible conflicts of interest
- professional indemnity insurance cover
- terms and conditions of appointment.

The Surveyor's duties of care

The professional duty

The professional objective of any chartered surveyor is to provide their client with timely and competent professional services within the guidelines issued by, and to the standards expected of, the Royal Institution of Chartered Surveyors (RICS). All practicing surveyors should therefore be familiar with recommended good practices and the minimum service standards set out in the relevant RICS Guidance Notes and Codes of Practice.

In the case of dilapidations services, the practitioner should at the very least be familiar with the RICS Guidance Notes on the following.

- *Building Surveyor Services* (2008).
- *Dilapidations* (5th ed, 2008).
- *RICS Short Form of Consultant Appointment* (2008).

In addition, Regulation 4 of both the RICS *Rules of Conduct for Firms* (2007) and the *Rules of Conduct for Members* (2007) require that professional work is carried out "with due skill, care and diligence ..."

The statutory duty

However, what is often forgotten is that the provision of professional services by the surveyor is also subject to Part II of the Supply of Goods and Services Act 1982. Consequentially, the surveyor is under a statute-implied term of appointment that the surveyor will provide their service with "reasonable care and skill".

There is an absolute duty on a surveyor to act objectively and competently throughout their appointment. The client, in making an appointment, will have a right to rely on the surveyor's duty of care. This will include relying on advice provided by the surveyor during the appointment and the right to rely, in good faith, on the contents of any claim documents or schedules prepared by the surveyor.

The ethical duty

In addition to the statutory and professional duty of care outlined above, there is also said to be an "ethical duty of care" for chartered surveyors when providing professional services. The definition of professional ethics endorsed by the RICS in the *Professional Ethics Guidance Note* (2002) is the:

> giving of one's best to ensure that clients' interests are properly cared for, but in doing so the wider public interest is also recognised and respected ...
>
> The need for professional ethics is based upon the vulnerability of others. The client (and where relevant, opposing negotiators) must be protected

> from exploitation in a situation in which they are unable to protect themselves because they lack the relevant knowledge to do so ...

The ethical duty owed by a surveyor, clearly extends to giving objective and competent advice and guidance to a client and, where necessary, having the courage to inform, educate and guide the client away from pursing a course of action that is inappropriate, unreasonable or that will expose them to unnecessary risk.

A good comprehension of the ethical duty of care will normally keep the surveyor attentive to the claim issues and associated risks. It should also remind the surveyor to avoid making unsustainable claims or defences.

Surveyor independence and conflicts

The handling of a dilapidations claim and the conduct of the parties is likely to come under close scrutiny where a conflict of interest exists or could be considered to exist. So that potentially damaging and costly disputes are avoided, a surveyor should carefully consider what (if any) possible conflicts of interest could exist before accepting any instruction.

Courts expect that conflicts of interest (potential or actual) are disclosed at the earliest possible point in the appointment process, so a party to a dispute can consider making alternative non-conflicting professional appointments.

The independence and reliability of the surveyor should also not be prejudiced by the surveyor engaging in what could be viewed as overly partisan negotiations or conduct.

Professional Indemnity Insurance considerations

Litigated dilapidations claims can vary in size and complexity, running into millions of pounds.

Given the complexities of dilapidations proceedings, the risks of breaching the duty of care owed to a client during the provision of services are significant. Therefore, the surveyor should remember that if they breach their professional duty of care and the claim or defence fails as a consequence, then the surveyor can expect to have a claim made against their Professional Indemnity Insurance policy.

It is a prudent measure for a surveyor to consider the possible size and nature of a claim and their own ability to provide independent and professionally competent services before they accept an instruction. As part of this process, the surveyor should consider the worst case scenario and satisfy themselves that they hold adequate insurance to safeguard their client's interests should the claim or defence based on their services fail.

If the appraisal of the insurance risk exceeds the level of cover held by the surveyor, then the surveyor should reconsider accepting the instruction.

Taking instructions

Instructions can be received from the party to the dispute direct, ie the tenant or the landlord. Instructions can also be received via lawyers, accountants or another surveyor.

Because the responsibilities of a dilapidations instruction can run for many years and ultimately result in the surveyor giving evidence in court, it is vital that the creation of the contract between client and surveyor is unambiguous.

The RICS Guidance Note, *Appointing a Building Surveyor — A Guide for Clients and Surveyors* (2001) suggests that an appointment agreement should be produced between the client and surveyor in which the brief, the scope of service and the conditions of engagement are clearly set out and agreed in writing. So when submitting service and fee proposals, the surveyor should ensure that all necessary and relevant aspects of the brief and terms and conditions of engagement are provided to the client.

It is also important to explain to the landlord or tenant client that the surveyor's overriding duty will be to be honest and truthful in their representations; and to conduct the claim in the manner expected by the courts and the Civil Procedure Rules (CPR) from the outset (see Chapter 4). The surveyor should explain that the landlord's claim will be restricted by the factors outside of their control, such as common law principles, statutory limitations and local market factors that influence the demand for a given property at a particular point in time. This approach will help manage the client's expectations in terms of their ability to manipulate their appointed surveyor.

Ideally, the surveyor should issue a detailed fee proposal that should be responded to in writing by the client in open terms with a signed letter of instruction.

Confirming appointments

The point at which the surveyor receives and accepts an instruction defines the point in time that the professional service responsibilities commence. The acceptance of an instruction should always be confirmed in writing by the surveyor.

It should also be remembered that should the dilapidations claim subsequently be referred to court, the terms and conditions of engagement and appointment brief will be disclosable documents. Consequentially, any untoward instructions or prejudicial content in the written appointment correspondence may undermine the reasonableness or appropriateness of the claim and action.

Part 2: Due diligence checks

Given the surveyors professional, statutory and ethical duties of care, all surveyors should seek to undertake initial due diligence reviews of issues such as:

- checking that they have the full set of relevant lease documents
- reviewing the validity of the lease
- undertaking lease "privity of contract" checks
- establishing client intentions
- obtaining other materially relevant information.

Failure to undertake appropriate due diligence on materially important issues can often have serious consequences for the claim. The due diligence checks covered in this chapter are not exhaustive and the surveyor may need to undertake additional investigations according to the circumstances of each instruction.

Obtaining and checking the lease

Obtaining the lease(s)

The RICS *Dilapidations Guidance Note* suggests that it is best practice for surveyors acting on dilapidations claims to "obtain a copy of the relevant lease" and to "satisfy himself that the documentation obtained is sufficient for him to discharge his instructions".

From the initial information gained while taking the instruction brief, the surveyor should be able to work out what lease documentation may be required or available. For example, where the client is the landlord and holds the freehold title interest and where there are no sub-tenancies, then there should only be a single lease document to obtain and review (although there may still be other additional tenure documents to be obtained).

Alternatively, where the client is the landlord or tenant in a lease with a superior landlord interest, a head tenant, and/or possibly multiple sub-tenants, then it is advisable for the surveyor to obtain copies of all the related documents so that the full context of landlord and tenant relationships can be understood and the various obligations appraised.

Once the extent of lease documents has been established, the surveyor should ensure that full copies are obtained in a format suitable for review and reproduction within the dilapidations claim.

It is always worth bearing in mind that copies of the lease(s) may also have been deposited with HM Land Registry (HMLR). Often, the HMLR copy of the lease is available through their online title search service in a downloadable PDF electronic format for around £10 per document. It may therefore be possible to obtain a complete copy of a missing lease from HMLR when other avenues have been exhausted.

Where it proves impossible to obtain a copy of the completed lease, the client should be advised that the dilapidations documentation will be prepared in good faith on the basis of the copy documents that are available, but that there are risks associated in doing so. If proceeding on this basis, it should also be clearly stated within the schedule and claim documentation from the outset. The tenant party should also be invited to make available any copies of the completed lease(s) that they may hold so that the schedule and claim can then be reviewed and revised as appropriate.

Checking the validity of the lease

The RICS *Dilapidations Guidance Note* suggests that it is best practice for surveyors acting on dilapidations claims to "obtain a copy of the relevant lease in complete form". However, what the guidance omits to clarify is what the "complete form" of a lease means. In order to understand this aspect better, one needs to understand the statutory requirements for a "complete" lease.

Under section 2 of the Law of Property (Miscellaneous Provisions) (LMPA) Act 1989, a contract for the sale or other disposition of an "interest in land" can only be made in writing. Furthermore, the contract must incorporate all the terms which the parties have expressly agreed, either by being set out within it; and/or by reference to some other document. The contract (lease) document incorporating the terms must also be signed by or on behalf of each party to the contract; although where contracts are "exchanged" the signatures do not necessarily need to be on the same copy of the contract document.

The granting of the lease results in the "other disposition" of an interest in a property and/or land for a temporary duration (the lease term). Consequentially, for a lease to be considered valid or in "complete form", it must be in writing, clearly sets out the terms, and have been agreed/signed by the parties.

As any dilapidations claim will be substantially restricted and prone to collapse if the lease does not satisfy the legal requirements of section 2 of the LMPA 1989, lease documents received should be reviewed to ensure that the version received is the final version of the completed lease document. For example, is it a draft version? Has the lease been signed by both parties? Is it dated? Does it carry the registration duty paid stamps, etc? If so, then it is unlikely that the express terms will have been changed, unless a deed of variation has been subsequently agreed.

It is essential that wherever possible the full and completed copy of the lease is obtained, as without it the surveyor has no way of knowing if there were any last minute variations or hand annotated amendments made to the lease during completion that may have a material bearing on the claim.

When undertaking checks, it is not uncommon for surveyors to find that the first versions of documents received are incomplete in some form or another. Often, this is because a client (or their managing agent) has never been provided with a full copy of the completed lease document by their solicitor and that they only hold a draft copy of the lease on their files, originating from the letting negotiation proceedings. It is also not unusual to discover that, eg the client may have changed solicitors since the granting of the lease and the original completed copy of the lease is missing, in archives or, in some instances, may have been destroyed.

Where there are concerns over the status of a lease, then further enquiries should be made with the client and/or their legal adviser to clarify the status of the documents. The consequences of a surveyor failing to carry out the most basic critical appraisal of the documents received may result in erroneous dilapidations claims being prepared and made in a reckless manner.

Lease privity and status checks

The importance of privity checks

Due to the fact that in the majority of cases a dilapidations claim is simply no more than a claim for breach of contract made under contract law, a dilapidations claim will normally only be possible where a valid contract exists between the two parties to the dispute.

Where no contractual relationship exists between the parties, the the landlord may find that they are dealing with a tenant at will, tenant at sufferance or similar. The landlord may also find that their ability to make any claim will be appreciably limited to the extent reasonably permissible under the alternative law of waste.

While taking instructions and establishing the brief, clients will normally inform the surveyor that "Company A" or "Mr B" is the tenant under the lease; or that they are the landlord, etc. The client may have made these representations on an entirely innocent and genuinely believed basis, but clients can often be mistaken.

Often, the lease may have been granted many years prior to the contemplated dilapidations action. The longer the period since the granting of the lease, the greater the potential for events having occurred that could affect the privity of contract.

For example, there may have been a licence to assign the tenant's interest that was not subsequently completed (eg no deed of assignment completed), thereby possibly rendering the licence "void". Alternatively, a tenant party may have gone into administration or have been "dissolved" or "struck off" and a new legally separate company may have been formed which thereafter continued with the occupation of the demised premises and the original tenant's "business as usual" without the landlord realising.

As the privity of contract issue is of material significance, all representations as to landlord or tenant party identities should not be taken simply at face value. Surveyors must ensure that they consider and understand the chain of privity of their clients lease.

Check who is paying the rent

One of the simplest privity and due diligence checks to undertake is to establish who is actually paying the rent under the lease. This sounds like an absurdly obvious check that any landlord or their managing agent would routinely undertake, but experience suggests that this is not the case. It is a surprisingly common problem to find that a landlord has not stopped to check who is making the quarterly payments of rent just so long as the rent payments are received each quarter.

The surveyor should try and establish the name of the corporate body or individual who has been paying the rent. This check may require the client to make enquiries with their bank if they do not hold sufficient details or copies of payments on their files. Where payments are found to have been received from a party other than the valid or known tenant, then it is likely that further enquiries will need to be made with the various "tenant" parties to establish the nature (if any) of the relationship between them.

Checking HMLR registered interests

The freeholder/superior landlord legal interest in any particular property can normally be quickly checked via an internet search of the HMLR register. The HMLR online register can currently be accessed via their website at *www.landregisteronline.gov.uk/*. The register also often holds details of registered leasehold interests and other materially useful/relevant documents.

While not all titles and interests in land have been registered with HMLR, in most cases the surveyor will be able to obtain copies (paper or digital format) of the registered title, the title plan and possibly even leases and conveyance deeds or plans, etc. The small cost of typically £3 to £6 and 10 minutes additional work per HMLR online title search, is so slight in the context of most cases that it should be routinely undertaken at the outset (and it may well even prove invaluable).

Each HMLR registered title will state the name and nature (eg freehold or leasehold) of the registered title. It may also list restrictive covenants or obligations that could be of interest and may allow further claims if they have been breached. A title may also list details of other known superior or subservient leasehold title interests or even previously unknown deeds of variation. This information may therefore help the surveyor to track down and obtain copies of all relevant tenancy documents.

If discrepancies are found between the named landlord and tenant parties described in the lease and the HMLR title(s), then the discrepancies should be raised with the client and their legal advisor and further clarification sought. It is likely that further investigations will then be necessary to establish if there are any known deeds of assignment, underleases or other transfers of legal title (such as third party company mergers or acquisitions of tenant parties).

Reviewing corporate identities

If the landlord or tenant is a UK limited company, they should be registered at Companies House. The Companies House online register search facility (*http://wck2.companieshouse.gov.uk*) may often reveal useful information on the identity of the tenant.

If checking a corporate landlord or tenant registry entry at Companies House, it is important to remember that a company has the ability to change its registered name. As a consequence, care should be taken to check that a search for the tenant's company produces a current registry entry for the company that had the tenant's registered name at the date of the lease or assignment, etc.

For example, say the tenant at the start of the lease was called XYZ Ltd. If the current registry search finds an XYZ Ltd but records show that it used to be called, say, ABC Ltd at the material dates of the granting of a lease or an assignment, then a search of the "previous names" or "dissolved names" registers may need to be conducted to find out what happened to the original tenant XYZ Ltd (and if still trading, what name it is currently registered under).

Checking corporate status

Surveyors can purchase copies of recent company reports directly from the Companies House website (see above) for a very small fee (typically £1 per document) and a further 10 minutes effort. Experience shows that corporate status searches frequently have a material bearing on the prospects of a claim.

Sometimes a search will reveal that a head or sub-tenant party has been dissolved or struck off and this should raise immediate questions as to the true identity of the "tenant" party with whom the landlord has been dealing and accepting rent. If the tenant no longer exists, then the prospects for a claim disappear and it may then become necessary to consider action against a guarantor (where they exist).

Even where the tenant party remains a going concern, the accounts information available from the Companies House online service may be of interest and may help establish the current covenant strength of the tenant and their ability to meet the cost of any claim.

Private tenant status checks

Where the tenant in question is other than a corporate body (such as a private individual) and if there are any concerns over the tenant's covenant strength (such as a bad track record in paying rent or current rent arrears), then it may be worthwhile undertaking some simple desktop diligence checks on the tenant.

Where checks on a private tenant are felt necessary, in addition to any address for a tenant recorded within the lease, the surveyor should find out the current correspondence address for the tenant. The HMLR titles for the known tenant's lease and correspondence addresses can then be checked to try and establish if the tenant has any (or sufficient) assets to either fund dilapidations remedial works or to settle a damages claim.

If a surveyor finds that any tenant has negligible worth, no assets or, worse still, sizeable debts, the surveyor can at least advise their client on the findings so they can review their position. If the prospects for success of a claim look bleak, the surveyor should give their client an opportunity to terminate their appointments and mitigate possible wasted or unrecoverable costs before they are incurred.

Dealing with privity discrepancies

When a discrepancy in the privity of the lease contract is found, it does not necessarily mean that there has been a break in the chain of the privity of contract or an unauthorised assignment or parting with possession of the property by the tenant. This is because there are statute-permitted circumstances where other "tenant" parties may legitimately enjoy the benefit of the lease contract without formal deeds of assignment being completed.

Section 42 of the Landlord and Tenant Act (L&TA) 1954 (as amended), provides that a "tenant" or the "landlord" named in the lease is now also deemed to include:

(a) a company in occupation in which the person named in the lease has a controlling interest; and
(b) a person in occupation who has a controlling interest in the company named in the lease.

Where a tenant identity or privity discrepancy has been found, it will therefore be necessary for further checks to be undertaken to establish what, if any, connection there is between the true tenant under the lease and the possible third party.

Where further checks fail to resolve any privity discrepancy, then a notice requesting the relevant information and clarification can be served on the last known true/formal tenant under section 40 of the L&TA 1954. The tenant will then be under a duty to provide the landlord with the requested information within one month of service of the section 40 notice.

Hopefully, the various privity check measures will resolve any discrepancy. However, if there remains uncertainty over the tenant status of the occupant or it is established that there is a tenant at will in the property, then the surveyor should seek further advice from their client's legal advisers on how to proceed.

Checking for other tenancy documents

Deeds, licences and schedules

Once the surveyor has received copies of the relevant lease(s), they should then carefully review the contents for references to other material tenancy documents such as other leases; coloured and/or annotated plans; HMLR titles covenant obligations; schedules of condition; schedules of works; schedules of landlord's fixtures and fittings, etc. If references are found to other documents, then the surveyor should request copies of the documents for further review. All relevant documents should also be reviewed to see if they are complete (no missing pages) and legible.

In addition, the surveyor needs to consider the possibility that other separate documents may not be listed within the lease and yet may have a material bearing on the claim or action. For example, have there been any deeds of variation, deeds of assignment or licences to alter granted during the lease term? It is also worth checking if there is a rent deposit deed as there may be additional clauses within the deed that may allow deductions of the landlord's costs from the deposit in the event of a tenant's breaches.

Pre-letting negotiations or side letters

In the first instance and more often than not, a modern institution lease will contain a clause stating that the lease represents the full understanding and extent of agreement between the lease parties. So, when appraising the matrix of fact with regards to interpreting the lease, the surveyor in most cases should be able to rely predominantly upon the contents of the lease.

However, this is not a 100% guaranteed safe approach for surveyors to adopt. Even with the existence of such a "full understanding" clause, it is not impossible for the lease to have been drafted in such a manner that it did not fairly represent the intentions of the parties at the time for entering into the lease. In circumstances where the lease does not fully or adequately reflect the intentions of the parties at the granting of the lease, it may be possible for one or other lease party to seek "rectification" of the lease via the courts. However, the courts are generally reluctant to interfere with the terms of a lease once completed and will normally only rectify the lease where there is compelling evidence that a mistake or error in drafting has occurred.

Before finalising a pre-survey appraisal of the lease and tenant's covenants or obligations, it is a prudent measure for the surveyor to check if their client is aware of any pre-letting negotiation agreements that may have a bearing on the claim. Similarly, enquiries should be made as to the existence of any side letters or agreements that may also influence the claim. Such pre-letting or side agreements may concern waiving of requirements to gain the landlord's consent for certain types of alteration or rights to use other (common) parts of the property outside the demise, etc. There may also have been more detailed agreements over what constitutes common parts or the extent of the demise.

Where materially influencing pre-letting agreements are found and where they contradict the express terms, conditions, rights, exceptions or obligations of a lease, the surveyor should seek to inform their client of any contradictions and seek client directions or their legal advisor's opinion on which interpretation of covenants and obligations should be taken into account when preparing the claim.

The landlord's intentions

The landlord's intentions for the property

An issue of paramount importance in dilapidations is establishing the landlord's true intentions for the subject property and/or land. This clarification of true intentions will form the cornerstone of the scope of service required by the client; will have implications for the remedies available to or sought by the client; and will ultimately dictate the claim and service strategy.

At the commencement of providing professional services, the landlord should be asked to disclose their future intentions for the property. The landlord should be encouraged to disclose any intentions they are firmly committed to and also any further "options" they are considering but have yet to commit to.

A surveyor exercising the ethical duty of care should be prepared to test, politely challenge and critically appraise disclosed client intentions for the property. If having sought to identify the landlord's intentions in an open manner and in circumstances that require the landlord to disclose their true intentions, the surveyor should then be able to proceed with their professional services in good faith on the basis of the intentions stated.

However, where there are any discrepancies in the intentions information gained or remaining doubts over the genuineness of the landlord's stated intentions, they should be tactfully and politely raised with the landlord. Further guidance should be provided to the landlord over the importance of the issue of intentions and the risks arising from misrepresentation that may ultimately influence the outcome of the claim.

Identifying the landlord's dilapidations objective

Before considering making a dilapidations claim, it is essential to identify the nature of the tenancy (see Chapter 5).

Once the nature of the tenancy has been established, the surveyor should consider the extent of available remedies to the landlord (see Chapter 6). When advising on appropriate remedies, the surveyor should take care to consider both the landlord's intentions for the property and the landlord's preferred remedy or remedies. They should then seek to formulate and get the landlord's approval for an appropriate claim strategy that will allow the landlord to successfully prosecute the claim and achieve the remedy objective.

Identifying the type of schedule required

At the point in time when the landlord instructs the surveyor to prepare the schedule of dilapidation, it is vital that the surveyor and client discuss and agree the required format of the schedule. This often also needs input from the lawyer involved. The onus is on the surveyor to be certain of the nature of the required documents before the schedule is produced, to ensure that the resulting post inspection schedule and claim documents are fit for purpose.

Other useful diligence checks

The surveyor may wish to undertake a few additional desktop diligence studies in order to gain a better understanding of the subject property and the possible factors that may affect the claim.

Section 25 or section 26 notices

It is worth enquiring if there have been any notices exchanged between the landlord and tenant parties under sections 25 or 26 of the Landlord and Tenant Act 1954, and if so, to obtain copies of the notices for review in case further information is contained in the landlord's notices or responses concerning their future intentions.

Property marketing reports

Where the property is being marketed as being for sale or to let, then the surveyor should seek copies of the agent's marketing reports and details.

Local authority applications

The surveyor should also undertake desktop studies to ascertain if there have been any recent decided or live applications for planning, listed building or conservation area consents; or for local authority building control approval, etc.

Registered address and occupants

If the surveyor needs to clarify any part of the subject's property address or wishes to check the identity of occupants registered to the address (say for reporting purposes or for undertaking other diligence checks), then a search can be undertaken via the Royal Mail website at *http://postcode.royalmail.com*.

The Royal Mail website address search facility will allow searches by either address or postcode and the search results will list the names of companies or individuals registered at the address. The details of registered companies or individuals at a property can help the surveyor check if there may have been any under-letting or parting of possession that may influence the claim strategy and/or the remedy to be sought.

Business rates register

If the property is a commercial property, then there may be some useful information such as property net internal areas and outline schedules of accommodation to be found in the Valuation Office ratings lists (see *www.voa.gov.uk/business_rates/*). The rating lists information may help the surveyor better understand the extent of subject property so the surveyor can plan their survey in advance.

Site maps and images

It may also be of assistance if the surveyor views aerial images of the subject property through online websites and services, such as Google Earth (*http://earth.google.com/*); or MSN Live Search maps (*http://maps.live.com/*). The aerial images for a site may help the surveyor understand site-specific issues, such as access restrictions or may provide a useful (fairly recent) image of the roofs, etc.

Due diligence checklist

Whether they are acting for the landlord or tenant, the surveyor should complete the necessary "due diligence" checks detailed below. It is also worthwhile ensuring that the surveyors and lawyers representing other parties in the dispute have been equally thorough. Mistakes and issues identified early on in a dispute tend to be cheaper to rectify than those that surface later on.

Appointment considerations

- ☐ Conflict of interest checked? (and declared if known/possible conflicts).
- ☐ Personal Indemnity insurance cover sufficient?
- ☐ Appointment terms and conditions quoted?
- ☐ Client appointment confirmed?

Obtaining and checking the lease

- ☐ Lease documents identified and obtained?
- ☐ Lease documents checked for validity?

Lease privity and status checks

- ☐ Identity of rent-paying party checked?
- ☐ HMLR registry entries checked?
- ☐ Corporate identification checked at Companies House?
- ☐ Corporate status checked?
- ☐ Private tenant status checked?

Checking other tenancy documents

- ☐ Deeds of variation?
- ☐ Rent deposit deeds?
- ☐ Licences to alter?
- ☐ Licences to assign?
- ☐ Schedules of condition?
- ☐ Schedules of fixtures and fittings?
- ☐ Pre-letting negotiations or side letters?
- ☐ Other. . . . .
- ☐ Documents obtained reviewed?

Establishing the landlord's intentions

- ☐ Has the landlord clarified their intentions for the property?
- ☐ Has the landlord provided a statement of intent?
- ☐ Agreed landlord's dilapidations objective?
- ☐ Agreed the type of schedule required?

Other useful diligence checks

- ☐ Obtained copies of section 25 or section 26 notices?
- ☐ Obtained agent's marketing report?
- ☐ Checked for local authority applications?
- ☐ Checked post office-registered occupants?
- ☐ Checked business rates register?
- ☐ Checked site maps and images?

8 The Dilapidations Survey

Part 1: Pre-survey preparation

When it comes to dilapidations, surveyors should be aware that service to be provided is not simply an exercise of visiting a property and producing a bulging schedule, listing every conceivable tenant breach, down to the most minute blemish. The inspection and scheduling of dilapidation is more subtle and involved than many think and there is a lot of truth in the saying that "proper preparation and planning prevents poor performance".

The dilapidations survey stage will require the surveyor to form their professional opinion on a variety of breaches where their opinion should be tempered by common sense; a fair interpretation of lease covenants; and by having due regard to the degree, reasonableness and proportionality of any necessary remedy; all within the context of the landlord's true intentions for the property.

The appropriateness of the survey and the future viability and success of any dilapidations claim will, therefore, be heavily dependant on how carefully the surveyor prepares for the survey.

Review and interpretation of the lease

Before undertaking an inspection, a surveyor will need to take time to understand the nature and extent of the tenant's obligations under the lease and any superior lease. In dilapidations claims, the lease clauses that more often than not have the greatest influence on the potential claim are the clauses dealing with the tenant's:

- extent of demised premises
- repair obligations
- decoration obligations
- alterations and additions obligations
- reinstatement obligations
- legal compliance obligations.

In addition to the key dilapidations aspects, there will normally be other probably materially relevant clauses dealing with issues such as compliance with statute; professional fees; use of the property; and/or schedules of condition or similar attached documents.

Determining the extent of the demise

An early aspect for the surveyor to review and establish is the extent of the property and land and/or outbuildings, etc that has been demised to the tenant under the lease.

In most lease documents, there will generally be a brief description of the extent of the demise in the particulars of the lease at the start of the document. This description may be contained within the main text as is common with old leases, or may be expressly defined within a "definitions" section of the lease.

Alternatively, a brief summary description of the demise within the particulars may simply make reference to a more detailed description contained within a "schedule" towards the end of the lease (typically the "first schedule").

Once located, the description of the demise should be checked to see whether it extends to the whole property or to only part; and if only part, careful note should be taken of the parts demised.

It is also often useful if the demised premises is then considered in context with the immediate local area and a quick comparison with Her Majesty's Land Registry (HMLR) title plan for the property will normally prove useful, particularly if the description is unclear as to the extent of land or structures.

Interpreting a tenant's obligations

The issues of correct and careful interpretation of the lease covenants is something that poses many problems for surveyors as there will normally be a degree of reasonable interpretation of a particular lease clause that may allow differing professional opinion to exist. However and when in doubt, the surveyor should seek to retain a reasonable approach to issues of interpretations and have regard to the accepted common law guidance on interpretation (see Chapter 14).

The surveyor and lawyer involved in a dilapidations claim should always read the lease and understand all obligations that the tenant may have in relation to repair, reinstatement and redecoration. There will also potentially be other direct covenants, eg to comply with statute, pay mesne profits, etc.

Schedules of dilapidations produced by landlord surveyors can mistakenly include allegations of breaches of covenant which are not the tenant's obligation. It is vital to check the lease, the lease plan and review the extent of relevant covenants.

Repairing and redecoration covenants

If a lease for a commercial property does not contain an express covenant in relation to repairing, a court will not assume that one exists.

The leading case of *Proudfoot* v *Hart* (1890) LR 25 QBD 42 defined the nature of repair obligation "to keep in repair" by confirming that such a clause imposes a duty to put a dilapidated building into repair in order to keep it in repair, even if it had never been in that state during the lease.

The principle of the tenant's repair obligation may also apply to additions or alterations to the premises during the term.

Where the lease contains a tenant covenant to "put and keep the premises in good and substantial repair and condition", it should be noted that this covenant is more onerous than a standard repairing covenant that might just require the tenant to "to put and keep in good and substantial repair" and which would not require the tenant to put the property into perfect repair (see *Proudfoot* v *Hart* (1890) LR 25 QBD 42) or into pristine condition (see *Commercial Union Life Assurance Co* v *Label Ink Ltd* [2001] L&TR 29).

Alteration and addition covenants

Most leases will normally contain clauses that will restrict or even prohibit the tenant from undertaking certain works of alteration or addition to the demised property without first having gained the landlord's consent (consent cannot be unreasonably withheld). The main purpose of such clauses is to prevent the tenant from carrying out works that may prejudicially affect the structural integrity or configuration of the property, to an extent where the tenant's works cause damage to the landlord's reversionary interests.

The extent of restriction and/or expressly permitted works that do not require consent will vary from lease to lease and so the clause(s), once located within the lease, will warrant close scrutiny. For example, it may be found that the erecting of demountable partitions by the tenant does not require consent or that non-structural alterations are permitted, etc.

The surveyor should have established the extent of any permitted alteration to the property during the initial due diligence checks. Where past landlord letters of consent for alterations or formal licence to alter deeds have been granted, then the consent documents will also need to be reviewed as there may be materially important additional clauses, covenants and obligations contained within the documents.

It is essential that surveyors endeavour to properly understand the extent of the alteration and addition covenants/obligations and known consents already granted. An appraisal and understanding of the extent of permitted or "authorised" alterations will help the surveyor to identify other "unauthorised" works during the on-site survey so that they can then be dealt with accordingly within the schedule of dilapidations.

Reinstatement obligations

In addition to alteration clauses dealing with the undertaking of tenant's works during the term, many leases also contain express "re-instatement" obligations or covenants requiring the removal of tenant's alterations and additions (possibly subject to conditions or a need for notice); and the restoration of the property to the pre-altered state.

Reinstatement covenants may be contained within the body of a lease or may also be found within the terms and conditions of any landlord's consent for tenant's works granted during the term and so any letter of landlord consent or licence to alter.

As many reinstatement clauses are conditional to some degree, they warrant close scrutiny, as there may be material conditions such as a requirement for the landlord to give, say, at least six months written notice before the end of the lease term before a tenant is obliged to undertake works of reinstatement.

Where a landlord's notice or request for reinstatement of alterations is required before a tenant becomes obliged to reinstate, then a schedule of dilapidations cannot allege a breach of reinstatement obligations or include a remedy claim/request seeking reinstatement, as the breach will not exist.

If conditional reinstatement obligations are not identified and the conditions properly executed, then the landlord may find that they are left with (and become the involuntary bailee of) unwanted tenant alterations/additions to the property at the end of the lease term which they then have to remove, probably at their own expense.

Legal compliance covenants

Within a modern institutional lease, there will probably also be a clause requiring the tenant to comply with acts of Parliament and subservient legislation including regulations, order, byelaws and recognised competent authority rules, etc. However, it should not be forgotten that even if the lease lacks a contractual "compliance" clause, the tenant will remain subject to their statutory duty to comply with the law in terms in any event.

Commonly, a lease "compliance" clause will consist of two parts, the first being a contractual obligation confirming that the tenant is to comply with statute during the lease term; and the second part generally being an obligation for the tenant to indemnify the landlord for any losses they may suffer in the event of a tenant's failure to comply, resulting in enforcement action upon the landlord.

Prior to undertaking a survey, the surveyor should consider whether or not the tenant is subject to statutory duties, such as maintaining building services in a safe and compliant state.

The tenant may also have employer, occupier, health, safety, welfare and/or user specific statutory obligations or liabilities that may require them to undertake repairs, maintenance or quite possibly even upgrades and improvements to the demised property, regardless of any contractual "repair" standards or limitations contained elsewhere within the lease.

Lease "compliance" clause and statutory compliance obligations are complex issues and are unfortunately common areas of dispute in dilapidations proceedings (see Chapter 14). Surveyors should therefore consider the extent of possible compliance issues carefully before undertaking a survey so they are familiar with the reasonable and legitimate compliance issues and will be able to identify breaches during their survey.

Other common tenant covenants

In addition to the core dilapidations covenants concerning repairs, decorations and alterations, most leases will also contain express covenants and obligations dealing with many other issues that will have a bearing on how the dilapidations matters may be attended to.

For example, there may be covenants that concern a landlord's "self-help" remedy for tenant breaches during the term ("*Jarvis* v *Harris*" clause). Such a clause could be of relevance to the surveyor when arranging access and discussing notice requirements, timings and enforcement remedy options with the client.

The lease will normally contain a "yield up" covenant describing how the tenant is expected to yield up (hand back) the demised premises at the end of the lease. Such clauses may help identify whether or not tenant fixture may (or even shall) remain in the property or may have other materially relevant additional obligations.

There are almost certainly going to be clauses in the lease clarifying what professional fees and costs, etc may be *reasonably* sought from the tenant; under what circumstances and whether or not these fees and costs, etc are to be payable on an indemnity basis. The surveyor should take these fees and costs covenants into consideration when undertaking services and ensure that the client is aware of any potential weaknesses or unreasonable aspects that may negatively affect the landlord's ability to recover costs.

There may also be a number of other complicating clauses and factors that all warrant careful review. Surveyors who skip or unduly rush the lease interpretation stage tend to find their carefully constructed claims have sometimes been built on weak foundations and may even collapse. Ultimately, the laborious process of diligently reviewing and interpreting the lease will enable the surveyor to undertake their survey and prepare the subsequent dilapidations claim with greater confidence.

Schedules of condition

Often when parties are negotiating a lease, the parties will look to protect and/or better define the tenant's future dilapidations obligations by attaching an agreed schedule of condition to the lease when the lease is granted.

The schedule will constitute a factual record, a standard of condition of the property at a given date (normally just prior to the granting of the lease). The lease clauses can then be drafted to expressly refer to the state of repair and/or condition evident within the schedule; and the tenant's obligations can be defined to require performance of the lease clause obligation (such as repair or decoration, etc) by comparison to the schedule.

However, where a schedule of condition is attached to a lease, it is often erroneously assumed that it is always the intention to define the tenant's maximum or upper limit obligations, although in practice the schedule can also be referenced in such a manner as to define the minimum standard instead. The schedule will only be of relevance to lease clauses that expressly refer to the schedule and it will have no material influence on any other clause.

If there is a schedule of condition attached to the lease, then the surveyor should carefully identify all of the clauses within the lease that expressly refer to the schedule. The surveyor should then ascertain whether or not the schedule sets a maximum or minimum limit of the lease clause(s) on a clause-by-clause basis; and whether there are any further tenant specific repairs or decoration obligations contained in other lease clauses that may override the limits imposed by the schedule references.

"Tenant-like" user obligations

It has been held by the courts that there is an implied obligation on a tenant under a lease, to use and occupy the demised property in a "husband-like" or "tenant-like" manner. This means that the tenant must not damage the property and should "do the little jobs about the place which a reasonable tenant would do". However, common law suggests that this does not extend to undertaking repairs, etc where no express repairing covenant exists.

However, the established leading precedents on tenant-like user are now over 50 years old and during that time the statutory obligations on tenants have significantly changed. Where there are express lease clauses concerning repair, decorations and compliance with statute, etc the implied tenant-like user will, in practice, be of little benefit to a landlord's dilapidations claim, but may be of assistance where there are no repair obligations. When considering the issue of tenant-like user obligations, the surveyor must have regard to the full "matrix of facts" and, in particular, the modern statutory "user" related obligations, including relevant statutory or "compliance" obligations (see above).

Considering reasonable market expectations

When considering the extent of a permitted claim that could be made against a tenant for breaches of the lease or for waste, it must be remembered that the remedy being sought is an equitable and civil remedy. In civil cases, the courts tend to favour claims that have made reasonable allowances.

The courts have held that the level of damages that can be recovered must be reasonable and proportionate to the loss suffered. The principles of reasonableness and proportionality apply even where remedial works are to be undertaken. For example, it would be unreasonable to claim a loss of, say, £50,000 incurred (or to be incurred) in actually undertaking remedial works if there was only, say, £20,000 damage to the reversionary interest if the property was left in a dilapidated state.

When considering the issue of reasonableness in dilapidations, the courts have repeatedly considered what a "reasonably minded" prospective (hypothetical) tenant or purchaser would expect of the property given its age, character, location, nature, use and local market conditions.

If the surveyor is unfamiliar with the local letting/sale market conditions for the subject property that exist(ed) at the material date of the claim, then the surveyor should consult with a local property marketing or sales agent and seek to clarify what the likely hypothetical tenant or purchaser expectations would be for the property. Also, if the property is being actively marketed, then the surveyor should seek to obtain a copy of the marketing report.

During the survey and subsequent claim stages, the surveyor should take into consideration the reasonable market expectations information gained; and seek to temper or adjust their determination of the reasonable remedial works necessary and any associated damage or loss accordingly.

The reasonable remedy period

If a terminal dilapidations schedule is required to be prepared close to the end of the lease (say within the last six months of the expected end of the term), then the surveyor should carefully consider whether or not it is reasonable to undertake the survey at such a late stage in the lease term.

The question of reasonableness in undertaking a survey should be the nature of the remedy to be sought and the reasonable time remaining available to the landlord under the lease to effectively enforce the chosen remedy. The surveyor should consider whether sufficient time remains within the term of the tenancy for:

- the survey to be undertaken
- the schedule of dilapidations and associated claim documents to be prepared (including third party reports where necessary)
- the claim to be served and the remedy sought under the appropriate notice
- the tenant to voluntarily and reasonably comply with any landlord's notice before the lease determines
- the landlord to take further remedial enforcement action in the event of continued tenant default after the expiry of the reasonable notice period and before the lease determines.

Where the time required for implementing the chosen remedy exceeds the time available for executing or enforcing the remedy, then any attempt to commence arguably futile or unreasonable proceedings will be open to challenge.

For example, if the landlord intends to serve a terminal schedule under cover of a notice to repair with only two months to go on the lease; and if the "self-help" clause of the lease requires the tenant to be afforded a minimum three-month notice period to undertake works following receipt of such a notice, then the landlord's action will fail as it was commenced too late.

In such circumstances, the tenant may argue that it was impossible for the landlord to implement any stated action or notice intentions and may even seek to have the notice set aside, alleging that the notice was unreasonably and recklessly served (see *Rous* v *Mitchell* [1991] 1 All ER 676).

If the landlord has left it too late to reasonably effect enforcement action or to implement a remedy, then they may find that their costs incurred are also contested and prove to be unrecoverable.

If the terminal dilapidations action is at risk due to timing difficulties, the landlord's surveyor should inform the landlord of the potential unrecoverable cost risk. The surveyor should seek the landlord's further instruction on whether they wish to proceed; or whether they wish to defer dilapidations action to the end of the tenancy and their only remaining effective remedy of seeking lease end damages.

Part 2: Conducting the survey

Professional survey standards

Any surveyor acting on a dilapidations claim should be professionally competent to do so. It is therefore expected that, at the very least, the surveyor is familiar with the Royal Institution of Chartered Surveyors (RICS) guidance notes on:

- *Building Surveys and Inspections of Commercial and Industrial Property* (3rd ed, 2005)
- *Building Surveys of Residential Property* (2nd ed, 2004)
- *Dilapidations* (5th ed, 2008)
- *Surveying Safely* (2nd ed, 2006).

The surveyor should also have regard for their employment-related health and safety obligations and those of the tenant. Consequently, if the tenant remains in occupation of the subject property, then the surveyor should request a site or property-specific safety briefing from the tenant upon arrival and seek to comply with any health and safety-related landlord or tenant imposed safety rules and regulations.

Mitigating survey costs

When undertaking the dilapidations survey, the surveyor should be ever mindful of the overriding objective of the Civil Procedure Rules (CPR).

The survey should therefore be conducted so far as is reasonably possible in a way that seeks to mitigate expense incurred; and that would keep expense and attendances incurred proportionate to the complexity and probable value of the potential dilapidation claim.

If dilapidations breaches are identified or suspected during the survey that warrant further investigation (such as intrusive investigations or specialist measuring or testing equipment), then a note should be recorded to this effect and arrangements made for return inspections and further investigations where it is considered proportionate and necessary to do so.

Where additional investigations expenditure is to be incurred, the landlord should be informed in advance, particularly as they will find themselves unable to recover any costs incurred if the investigations fail to confirm the presence of any suspected dilapidations and further breaches of the lease.

Where third party consultant costs are to be incurred, then in order to keep both parties' costs to a reasonable level, the surveyors for both the landlord and tenant should so far as possible seek to agree a suitable single joint appointment of a suitable third party consultant; and should obtain the consultant's report on an "impartial expert" basis so that it may be relied upon by both sides to the dispute.

Similarly, where specialist plant and equipment is necessary to undertake a complete survey, both landlord and tenant surveyors should arrange for specialist plant and equipment attendance on a mutually convenient return inspection date so that both surveyors may have equal access to; and use of the specialist equipment on site.

Notice for access

If the surveyor is acting for the landlord and is required to inspect the subject property before the lease has determined, then the surveyor should have due regard for the tenant's rights for quiet enjoyment under the lease.

The landlord's surveyor should review the terms of the lease to see if there are any specific requirements with regards to gaining access when the attendances are to be undertaken as a surveyor or agent for the landlord. Typically, the landlord's surveyor will be obliged to give the tenant reasonable advance written notice, save for where access is needed immediately in the event of an emergency. Unless the lease states otherwise, it is generally considered that advance written notice of one week should suffice.

A notice for access should inform the tenant of the professional appointment of a surveyor by the landlord and provide contact details for the appointed surveyor. It should also expressly refer to the relevant clause(s) within the lease under which there is a right for access and the request for access is being made.

Given the tenant's occupier and possibly employer liabilities to staff and/or visitors, the notice should make clear the date and time where reasonable access will be required and whether or not any third parties or specialist plant and/or access equipment will also be attending. If intrusive investigations are planned that may cause some localised damage or disturbance, then the tenant should be informed of investigation and making-good works intentions.

Given the surveyor's professional duty and obligations in relation to surveying safely, it is also reasonable for the surveyor to request the opportunity to review the property or site-specific safety information and files upon arrival (and to list any particular information which the surveyor wishes to view, such as electrical test certificates or asbestos registers, etc). This approach also provides the surveyor with a reasonable opportunity to check if the tenant is actually complying with their tenant-like user and/or statutory obligations.

Within the notice for access, it is also reasonable to inform the tenant of the possible costs consequences under the lease, should they fail to allow or provide access. However, where time is not of the essence and where it is reasonable to do so, the surveyor should seek to invite the tenant to make arrangements for access on an alternative date subject to the surveyor being given sufficient advance notice.

When issuing a notice for access to the tenant, the landlord's surveyor should endeavour to serve the notice formally and in keeping with section 196 of the Law of Property Act (1925) and the Recorded Delivery Service Act (1962).

If the surveyor fails to submit reasonable access notices in advance of their survey in an appropriate manner, then the tenant may reasonably refuse to provide access. The landlord may then find their requests for the surveyor's costs for aborted attendances are challenged and prove unrecoverable.

Undertaking the survey

Survey methodology

The surveyor should approach the undertaking of the survey in an appropriate and methodological manner, such as those recommended in the various RICS building survey guidance notes.

It is not reasonably practical to attend every property with every conceivable tool or piece of survey equipment that may be necessary to complete a survey. However, the surveyor should ensure that they are in possession of a reasonable set of tools and survey equipment to be able to undertake the survey as envisaged in the initial client project brief.

Upon arrival at the site, should a surveyor discover an aspect of the property that will require specialist tools, equipment or third party attendances, then they should record as much of the property as they can and make the necessary arrangements to return with appropriate equipment or third parties at a later date (see also survey costs issues below).

During the survey, the surveyor should seek to reasonably examine those accessible parts and make notes on areas where access was limited, restricted or impossible. The surveyor should seek to take sufficient notes or record observations, sketches and measurements, etc while on site that will allow a considered dilapidations claim to be prepared (or reviewed).

The surveyors should also seek to record a reasonable and comprehensive set of photographs, or even prepare a video record of the property and land in a manner that would allow cataloguing and indexing and presentation as evidence in support of a future court claim or defence. These should be indexed as soon as possible.

Surveying in context

While undertaking the survey, the surveyor should seek to maintain an independent and professional approach. The surveyor should not only seek to identify technical breaches of the express lease or tenancy covenant terms; but should seek to dispassionately and reasonably consider the severity and degree of breach. In doing so, the survey should consider both the extent of the breach and appropriate remedy having due regard to the tenant's leasehold obligations.

If there is an agreed schedule of condition attached to the lease that affects the tenant's dilapidations obligations, then the surveyor should pay particular attention to what reasonable remedy is available (or even possible) for deterioration in the premises condition that gives rise to an actionable breach.

If the surveyor is in doubt over the severity of a breach or an appropriate remedy that would be expected in the context of the "reasonable minded" hypothetical purchaser or tenant, then the technical breach should still be recorded but the identification of the appropriate remedy and associated damage or loss to the landlord (if any) may require further post-inspection consideration.

Where the landlord perhaps is contemplating potential redevelopment or refurbishment options for the property but has yet to firmly commit to the option (eg move beyond the "zone of contemplation"), then the surveyor should record the full extent of breaches and remedies in case the landlord subsequently abandons the redevelopment options and seeks to maintain the property in its current configuration and state.

While undertaking a survey of premises, the surveyor should also look at the immediate locality so the premises can be viewed in a wider local context. The surveyor should try and form a general opinion on the normal or typical standard of repair of adjacent or similar properties. If there are a number of properties locally that are for sale, for let, vacant or even derelict or boarded up, then it may be an indication that there is a difficult local real estate market.

The local comparative market information may help the surveyor then review and possibly reappraise their opinion on the reasonable hypothetical tenant or purchaser expectations. This may also help the surveyor to reappraise the reasonable or proportionate remedial works to be claimed or to better consider possible diminution limits to any claim under section 18(1) of the Landlord and Tenant Act 1927 and under common law.

Tenant alteration complications

During the survey, the surveyor should seek to critically examine the property as there is only so much information that can be gained and understood in the pre-survey desktop study stage. An aspect requiring particular careful review on site is the issue of possible (previously unknown) tenant alterations and additions to the premises. Pre-survey assumptions may therefore require review on site.

When considering possible alterations and additions, the surveyor should look out for materials, components, installations, structure, fabric and/or finishes that look out of place within the property or that do not feature on any plans for the premises contained within the lease or tenancy documents. Where suspected alterations or additions are found, they should be recorded and examined for dating evidence.

Dating evidence for alterations is often easy to find, as many mass-produced construction components will have manufacture date stamps or moulding marks. The dating evidence may prove particularly useful for establishing:

- when the alteration or addition took place
- if there has been more than one tenant during the tenancy, which tenant undertook the works
- whether the works, if constituting a breach of the lease or tenancy, remain actionable breaches or are time barred breaches under the Limitation Act 1980.

Where suspected tenant alterations and/or additions are discovered, further checks should be undertaken with both the landlord and tenant to ascertain if they were alterations or additions undertaken with or without consent; and when they were undertaken. It may also help to check with the local planning or building control authorities to see if they hold information that may help determine when suspected alterations were undertaken.

Where alterations or additions are found, the surveyor should also ascertain if the landlord wishes to take action for their removal; or if the landlord would be happy for the alteration or addition to remain subject to being yielded up in good repair and condition.

Where tenant's alterations and additions to the property remain in the demise after the end of the lease or tenancy, then the landlord should take further legal advice on their "involuntary bailee" rights and remedies; and the claim on the tenant (or their predecessor in title) responsible for the breach should be prepared and structured accordingly.

Recording survey evidence

The surveyor should remember that all details and information recorded on site will constitute disclosable evidence and so care should be taken to retain site inspection notes and/or records in their original form. The surveyor should also be prepared to make copies available upon request at an early stage in proceedings. Furthermore, the evidence recorded or obtained on site should be held in a suitable format so that it can be readily reviewed, examined and understood at a later stage.

Where it is necessary, proportionate and possible to do so, the surveyor should seek to annotate building/site plans or freehand sketches with sufficient information to supplement other records or evidence recorded and to assist the understanding of locations and features mentioned within the schedule or claim documents once prepared.

If recording a series or sequence of repetitive structures, fabric or features of similar appearance, it can be useful to individually label and record the similar features during the survey. Where undertaken, the serialised recording of features should be undertaken in a logical sequence. For example, windows could be listed starting with the front elevation, from left to right, top floor to ground and then working round the building in a clockwise direction. Any serialisation of features can be recorded on sketch plans or elevations and also in photographs (where possible) by using, say, handwritten serialised "post-it" notes fixed to the features so that they are discernable in photographic evidence.

If the surveyor intends to record a comprehensive photographic or video record, then it is important that sufficient information is recorded to allow the context of the photographs or video footage to be understood. This may mean taking a sequence of images that allow a specific defect to be viewed in both a wider context (say relative position on an elevation) and also in close-up detail.

The surveyor should also consider that there is potential in each claim for a protracted dispute and a court case to follow; and that the evidence may need to be reviewed and may even be cross-examined in the future, sometimes years after the survey was undertaken. Over time the surveyor's detailed memory of a survey will diminish and so a good surveyor tip for when recording image-based evidence is to introduce a pointer (finger, pen, stick, etc) or some form of scale into the image while recording it on site.

Introducing scale or pointers into imagery during the survey will help later reviewers of the evidence focus on the area of the image where the defect is located. It will also help subconsciously demonstrate to the reviewer that a defect was recorded and observed at the material time and with conscious thought by the surveyor (as opposed to the impression that can be had by a seemingly robotic, less considered sequence of "point-and-press" non specific imagery).

Finally, on the issue of evidence, given the developments in digital recording technology over the last 10 years and the continuing rate of technological progress, many surveyors are recording their notes, comments, observations and photographic evidence in a digital format. Digital evidence is proving a useful tool in claims, but the power of modern personal computers means that most computer users could in theory undertake substantial manipulation and distortion of the evidence and so care must be taken on how the digital evidence is recorded and subsequently held. Where surveyors choose to use digital evidence, they should have due regard to the requirements of the RICS guidance note *Electronic Document Storage — Legal Admissibility* (2nd ed, 2003).

Post-survey client review

Following the on site survey, the surveyor should seek to provide their client with some initial feedback and further guidance on their observations and opinions of the potential claim. They should also seek to advise of the necessity; and seek approval for any further attendances by the surveyor or third party consultants and contractors.

Where further attendances and investigations are approved, the surveyor should co-ordinate all further attendances and appointments to ensure that any third parties are also adequately briefed, as they may be unfamiliar with the rigours of attending to dilapidations claims.

The surveyor and their client should also review the original surveyors brief and claim remedy objective in case the on site findings suggest that an alternative remedy might be necessary or more appropriate.

For example, the original intentions might have been to prepare a schedule of dilapidations for service on a tenant under a notice to repair as the first step in "self-help" enforcement action. However, it could have been found on site that the property is so dilapidated that it is unsafe and poses a "dangerous structures" risk, in which case there are strong grounds to seek a more immediate remedy such as re-entry and determination of the lease.

When all immediate post survey and additional investigations have been satisfactorily concluded, then the surveyors should proceed to prepare and compile the schedule and any claim.

9 Preparation of the Schedule

Introduction

Reducing the costs of the schedule

Following the judgment in *British Westinghouse Electric & Manufacturing Co Ltd* v *Underground Electric Railways Co of London Ltd (No 2)* [1912] AC 673, it is a long established principle that a party to a damages claim has a duty to mitigate their loss, ie to take: "all reasonable steps to mitigate the loss consequent on the breach".

The duty to mitigate the loss applies to professional fees of surveyors preparing a schedule. Just because a lease allows for recovery of a surveyor's fees, does not mean that all fees are reasonable. Care should be taken not to incur unreasonable (and probably unrecoverable) costs.

For example, it is commonplace still for many surveyors to spend time reproducing the clause text within their schedules by re-typing the clauses. This is not only wasteful in terms of effort but is prone to introduce avoidable errors. It may also distort the lease by taking individual clauses out of context of the wider interpretation and intentions of the whole lease. It is therefore better practice, quicker and more cost effective to simply provide a clear and legible copy of the whole lease within the appendices of the schedule (in either paper or digital format).

By way of further example of unnecessary attendances, many surveyors spend time and effort in costing or pricing the breach remedies when the landlord is not seeking damages and may actually be seeking the self-help remedy under a notice to repair. In such circumstances, the surveyors "estimate" prices are entirely academic and irrelevant as either the landlord's or the tenant's tendering contractors will submit quote for the works in any event.

There are many other occasions where the approach to schedule preparation may result in unreasonable and unnecessary costs and the surveyors should consider their actions and endeavour to avoid wasted effort and unnecessary costs in keeping with the overriding objective of the Civil Procedure Rules.

The Civil Procedure Rules 1998

Given the overriding objective of the Civil Procedure Rules (CPR) requiring timely, effective and proportionate attendances on civil disputes, all dilapidations claims must be structured and processed so that both parties can clearly understand the nature and basis of the claim and the remedy being sought to resolve the dispute. To achieve this aim, claim documents should be prepared and set out to clarify:

- the particulars and basis of the claim
- the nature and extent of any alleged breach
- the nature, extent of any remedy allegedly required
- the reasonable timescale for responses and/or voluntary remedy
- a summary of claim for damages.

The above aspects are normally stated within the "particulars of claim" within a cover letter or notice and/or within the schedule of dilapidations document itself.

Where the remedy being sought does not involve a claim for financial damages in the event of continued tenant default (such as when issuing a "notice to repair" on a tenant during the lease term), then a schedule of dilapidations served upon the tenant with a suitably drafted "notice" should suffice.

Where the remedy being sought will include a claim for damages from the tenant in the event of their continued default, then the schedule of dilapidations and notice will need to identify the value of the damages. In such circumstances, a "summary of claim" should also accompany the claim documents.

The preambles

Because the schedule of dilapidations is a document prepared in contemplation of litigation, it has become increasingly accepted for a schedule to be capable of being read as a stand-alone document.

The schedule should include detailed "preambles" at the start of the schedule that present the claim clearly and enable all parties to understand the context in which the documents were prepared.

Parties to the dispute

It is vital that the schedule identifies clearly who is involved in the dispute, ie to identify the party claiming the damages and to identify the party from whom damages are sought. These may eventually be the claimant and defendant in a dispute and the judge and lawyers will benefit from a clear understanding from the outset.

The claim preambles should set out clearly who the parties to the claim are and what their respective landlord and tenant interests are in the subject property (see Chapters 5 and 7). The full names of the landlord and tenant should be set out clearly on all correspondence, notices and schedules.

Relevant tenancy documents

The preambles should clarify what tenancy documents have been taken into consideration and where possible, full copies should be provided with the claim documents when served (preferably within the schedule appendices — see below). This is so the schedule and claim documents can form a "stand-alone" package of information that allows any reader to ascertain the materially relevant aspects of the claim from one source.

Remedy sought

The preambles should seek to set out in plain and simple English the surveyor's understanding of landlord's desired remedy; and where possible their intentions for achieving or implementing the remedy and any reservations being made by the landlord to consider alternative remedies or action should the need arise.

Landlord intentions

The landlord's intention for the property is of paramount importance (see Chapter 2). A claimant landlord cannot recover what they have not lost and the only way of assessing the loss is to factor in landlord intent. For example, if the landlord has already agreed to let/alter the premises without doing the works and without suffering any loss of value or rent, etc then a claim for lease end breach of covenant will struggle. Equally, if the landlord intends to alter or demolish, then section 18 of the Landlord and Tenant Act 1927 applies.

Without knowledge of the landlord's genuine intentions for a property, dilapidations claims remain theoretical and claims become hypothetical exercises and are effectively incapable of proper resolution. Where intentions are unclear, any settlement negotiated will be at risk to one or another of the landlord and tenant parties.

Worse still, if a settlement is reached on the basis of reckless or knowingly false (fraudulent) representations of intentions, then the settlement could be overturned and re-opened even where it was reached on a "full and final basis" (see Chapter 14).

The preambles should make clear the landlord's intentions for the subject property. Where the property is being marketed for sale or re-letting, then details of the landlord's agents should be made available. It may also be necessary to arrange for disclosure of any sale or letting agent's advice to the landlord at a later stage in proceedings or upon request.

Where a landlord's intentions have already been openly declared (such as within a section 25 of the Landlord and Tenant Act 1954 notice), then it would be reasonable but not essential to refer to the previously notified intentions within the preambles.

Where the landlord has yet to commit to a firm intention beyond the mere "zone of contemplation", then the various future intention options should all be openly stated, together with the landlord's intended option review strategy and programme.

Where negotiations are based on one set of stated landlord intentions and these subsequently change before the claim is settled, all parties need to be informed of changed intentions in a timely and open manner so they can review and, where necessary, revise their position.

Appointed professional advisors

The preambles should make clear the nature and role of any professional advisor or surveyor appointments made in relation to the claim. The details should allow the individual surveyor who prepared the schedule to be identified, together with clarification of any assistance they may have received from colleagues.

The details provided should briefly summarise any professional qualifications the surveyor holds so their relevant expertise to attend on such matters can be ascertained. Contact details for the surveyor and their professional practice should also be made available.

For transparency in proceedings, the surveyor and other professional advisors should clarify the extent and basis of their engagement and clarify whether or not they have any possible or known conflicts of interest that could be considered to be prejudicial to future proceedings.

The surveyor and other appointed professionals should be prepared to make full copies of the terms and conditions of engagement available upon request.

Survey details

Information should be provided for dates of the attendances and inspections of the property upon which the schedule is based.

The surveyor should provide sufficient information for the reader of the schedule to understand the schedule contents and should include identification plans and sketches within the appendices where necessary (see below).

Where relevant and necessary, further clarification should be provided with regards to restrictions and limitations that may have restricted the extent and degree of inspection and/or services undertaken.

Where further inspections are/or may become necessary, then care should be taken to reserve the claimants rights to undertake the further inspections and revise the schedule and associated summary of claim as necessary.

Glossary of common terms

Where common terms and phrases are used throughout the schedule documents such as references to standards or states of condition, etc then it may be useful to provide a glossary of the common terms to aide review and understanding of the schedule.

The commonly used descriptive terms relating to condition of elements (such as "good repair") could for example be made on the basis of the following definitions:

Condition Expression	Definition
Good	In new condition with no soiling/wear or other defects.
Fair	Subject to general minor wear and tear with slight signs of soiling but generally still serviceable and functioning adequately.
Moderate	Subject to prolonged wear and tear, requiring attention and/or repairs although still generally serviceable.
Poor	In a deteriorated/dilapidated condition, subject to prolonged wear and tear and generally requiring substantial repairs and/or renovation.
Very poor	Extensively dilapidated condition requiring extensive and major repair and/or renovation.
Obsolete	Generally in such a condition that it is beyond economic repair and/or renovation and requires wholesale removal and rebuilding/replacement as necessary.

Also, all cracks discovered within the premises at whatever location have been defined in accordance with the Building Research Establishment standard as follows:

Crack	Definition
Minor crack	Less than 1mm wide.
Moderate crack	Greater than 1mm but less than 5mm wide.
Severe crack	Greater than 5mm wide.

Assumptions

The preambles should clarify issues such as any glossary definitions and assumptions that affect the understanding of the claim.

Assumptions such as the surveyor's consideration and opinion on the reasonable expectations of a hypothetical prospective purchaser or tenant should be made clear. Assumptions as to standards of remedial works materials, workmanship and procurement should also be stated.

Restrictions and limitations affecting the extent of survey should be clarified and where further investigations are planned, the plans and intentions in this regard should be made clear.

Where there is potential for further revisions and change to the schedule of dilapidations, then suitable reservations in favour of the landlord should be made allowing the schedule to be revised and reserved at a later date up until the point of a full and final settlement being agreed.

The schedule of dilapidations

Basic objectives

A schedule of dilapidations should be sufficient in terms of content to adequately describe the alleged breaches, the remedies, and where necessary and appropriate, the cost of performing the remedies.

Ideally, the schedule should also be structured in a logical manner that allows the contents to be easily followed by any reader and proportionately presented according to case-by-case specific circumstances.

Items "rendered valueless"

The "second limb" or part of section 18(1) of the Landlord and Tenant Act 1927 (L&TA 1927) prevents a claim for damages being made in relation to wants of repair or decorative repair where the remedial works would be "rendered valueless" or "superseded".

Consequently, when preparing a schedule of dilapidations, the surveyor must seek to take into consideration the possible restrictions on the schedule contents and claim imposed by section 18(1) of the L&TA 1927 (see Chapters 2 and 15). It is expected that surveyors omit, from the body of the schedule of dilapidations, any potential claim item that they know will be rendered valueless by the landlord or future tenant's alternative work intentions.

Basic schedule format

A typically encountered initial schedule of dilapidations format may look as follows:

Item	Lease Clause	Breach of Covenant	Remedy	Cost
1	3.42	The tenant has installed 5 l/m of base units and 4 l/m of wall units to form a kitchenette	Strip out base and wall units, remove from site, make good	£325
2	3.42	The tenant has tiled the splash-back area above the work top (5 l/m)	Remove tiling. Make good plastered surface ready for redecoration	£285

This common but basic example allows all the necessary information, record of breaches and requested remedies to be crisply and simply conveyed to the tenant so they know what is required of them to address their breaches and dilapidations; and so they may take measures to fully comply with notices received.

A typical schedule of dilapidations is broken down into sections covering:

- repair
- redecoration
- reinstatement of alterations
- statutory or regulatory compliance.

The "Scott" schedule format

Where the schedule of dilapidations claim is likely to be contested and become the subject of a landlord and tenant dispute that could result in litigation, it is considered better practice for the schedule to be presented within a more extensive "Scott" schedule format (named after a G A Scott who was an Official Referee in the 1920s and 1930s in the equivalent of the modern day Technology and Construction Court).

The alternative "Scott" schedule format is similar to the initial schedule format referred to earlier, but includes two additional columns for the landlord and tenant's surveyors to make comments or remarks in respectively. It also includes four columns in which each surveyor may cost/value both their own opinion of the breaches and remedy required; and the other surveyor's alternative opinions of breach and remedy.

The purpose of the Scott schedule is to enable the sides to a dispute to quickly and efficiently set out their opinions and views. It also enables a court to select the costs it thinks is appropriate on an item-by-item basis according to its judgment on each item where the combined sums of court-approved figures will be taken into consideration when awarding a total claim value.

It is now becoming increasingly common for a schedule of dilapidations to be served in a Scott schedule format from the outset of the claim. An example page form a Scott schedule can be found in the appendices of the Royal Institution of Chartered Surveyors (RICS) *Dilapidations Guidance Note* (5th ed) and as follows:

Schedule appendices

There are many documents relevant to the schedule of dilapidations (see over, p132). Documents should only be included where they add to the reader's understanding and clarify the claim. Useful appendices to include are:

- copies of the materially relevant tenancy documents
- copies of photographs
- useful supporting plans and sketches
- third party reports.

Full copies of tenancy documents

In an area of property dispute where significant issues can be determined by punctuation and grammatical nuances, it is very dangerous to summarise individual lease clauses relating to dilapidations claims. To a lesser extent it is also potentially problematic to extract a selection of lease clauses to be copied into the schedule.

It is far better to copy the entire lease and licence documentation to create a separate appendix to the schedule. This will provide one substantial bundle of documents for the negotiation and investigation stage of the claim, following service by the landlord's lawyer.

Item No	Clause No	Breach Complained Of	Remedial Works Required	Tenant's Comments	Landlord's Comments	Landlord's Item		Tenant's Item	
						Landlord's Costing	Tenant's Costing	Landlord's Costing	Tenant's Costing
2 EXTERIOR WANTS OF REPAIR, RENEWAL AND REINSTATEMENT									
2.1 ROOFS									
1	4.(3)	Numerous cracked, broken, slipped or missing terracotta roof tiles across the roof pitches. 1 No broken bonnet tile to the roof over the rear second floor dormer window.	Retain, re-use and re-fix all slipped terracotta roof tiles. Supply and fit replacement terracotta roof tiles where existing tiles are missing, cracked or broken. Supply and fit 1 No replacement bonnet tile to the second floor dormer window roof.			2,195			
2	4.(3)	1 No broken terracotta finial on the South East return at the junction between the front roof and original East side roof.	Strip off and replace 1 No terracotta clay finial with replacement globe headed terracotta finial to match original.			305			
3	4.(3)	3 No eroded and holed terracotta chimney pots to chimney stack "CH1".	Strip out and cart away 3 No terracotta chimney pots from chimney stack "CH1" and supply and fit replacement terracotta chimney pots to match original. Make good all disturbed haunchings.			732			
4	4.(3)	Cracked brickwork and missing/friable mortar to high levels of chimney stack "CH2". Friable and eroding pointing and mortar beds to ridges and chimney stacks.	Undertake localised epoxy resin crack bonding repairs using BBA accredited materials and to BBA accredited methods. Allow for the insertion of structural bars/ supports within masonry to chimney stack "CH2" to the Structural Engineers satisfaction and make good disturbed pointing and mortar on completion. Cut out and stitch in replacement bricks where existing brick faces have spalled with soft clay stocks to match existing from reclaimed sources. Rake out and reinstate cement mortar haunching and ridge beds where currently friable.			2,439			

Address
Rev #

Scott Schedule page # of ##

Photo schedules and indices

"A picture paints a thousand words". If lease end schedules included a clear photographic record of the state of the demise as it was at the date of the claim, then scope for future dispute is reduced and the claims process would improve.

If the surveyors are recording relatively few photographs then they can be individually numbered accompanied by brief descriptive text and the details of the recording location given on a plan of the demise. Within each item of the written schedule there can also be a separate column for the surveyor to insert a relevant photograph reference from the numbered photo schedule.

Alternatively, if the photographic (or video) record of the property is extensive (as is increasingly common in the modern digital age), then it may be better to simply catalogue the photographic record by location in folders on a CD ROM and to provide a text index and "thumbnail" catalogue prints.

However presented, the photographic evidence should form an integral part of the schedule and should allow any reader to readily review and understand the dilapidations supporting evidence. This record may prove to be invaluable and may be pivotal at later stages in a claim, for example at a point in time when the building may not be available to re-inspect, or when the building has been substantially changed.

Block identification plans and sketches

A block plan or sketch either with room/location/element identifying annotation or grid lines will provide an unambiguous means by which both landlord and tenant surveyors can refer to the locations of elements of the building.

For example, where available, the lease plan of the demise can easily be copied from the lease with hand drawn annotation added.

Third party reports

The building surveyor preparing the schedule will have expertise relating to the fabric of the building, the fixtures and various installations. When deciding whether to use an external specialist, the surveyor should ask themselves the following questions:

1. What will they add?
2. Are the specific issues beyond a building surveyor's area of expertise?
3. Are the specific issues substantial enough to warrant the extra cost (the issue of proportionality)?
4. Is the additional cost recoverable?
5. When instructing a third party expert the following points should be considered:
   (a) terms if appointment — ensure the letter of appointment clearly defines the scope of the instruction and reinforces the duty of the impartiality that the consultant will owe pursuant to the Civil Procedure Rules (CPR) 1998 in any written report. A third party specialist who reports and negotiates on a partisan basis will undermine the credibility of the appointing surveyor, their client and the claim/response document. This will ultimately delay settlement.
6. Who is appointing the third party?
7. Has all relevant history and documentation been issued to the third party, including a full description of the claim, a detailed summary of the relevant tenure documents and a summary of the form of report that is required?

An experienced third party brought on board to assist the surveyor in the preparation of the schedule will ideally be well versed in the format of the schedule and will be able to produce a clear and impartial list of breaches, remedies and costs (as appropriate) that the instructing surveyor can quickly include into the main schedule of dilapidations claim.

10 Collating the Claim

Claiming for the loss

What is loss?

In *Ruxley Electronics and Construction Ltd* v *Forsyth* [1995] EGCS 117, Lord Bridge of Harwich considered that:

> Since the law relating to damages for breach of contract has developed almost exclusively in a commercial context, these criteria normally proceed on the assumption that each contracting party's interest in the bargain was purely commercial and that the loss resulting from a breach of contract is measurable in purely economic terms. But this assumption may not always be appropriate.

Surveyors should appreciate that quantification of a loss is not necessarily a straightforward mathematical calculation based on the economic or commercial positions of the parties. It may also involve an assessment of all shades of risks and possibilities; and may involve an assessment of a loss of a chance, amenity or other losses. These factors consistently lay at the heart of leading case law judgments and should form the cornerstones of any surveyor's appraisal of loss.

When preparing a summary of claim for dilapidations damages, the surveyor will need to have due regard to the common law principles of loss (see the key principles summarised in Chapter 2).

The risk of false claims for loss

A current "hot topic" in the dilapidations area concerns the apparent willingness by a minority of dilapidations claimants/defendants and appointed professionals of all disciples to make rash, reckless or knowingly false claims or defences concerning the levels of loss allegedly suffered. This minority is often influenced by the desire to maximise the returns and potential profit associated with performance-related fees for their services. However, thankfully the Royal Institution of Chartered Surveyors (RICS) has recently started taking a more active approach in trying to halt these unprofessional and fraudulent practices and, in April 2008, launched a Fraud Response Plan. The recently revised RICS *Dilapidations Guidance Note* (5th ed, June 2008) also advises members that:

> Surveyors should not allow their professional standards to be compromised in order to advance clients' cases.

Where a claim for loss is to be made (or challenged during a defence representation), the parties making the statements should use their best endeavours to ensure that any representations they make are made honestly and with justification. Where claims or defences are made on any other basis, then the parties making the representations will be at risk of making false representations that could be considered to be either innocent, negligent or possibly even fraudulent (the latter of which may even constitute a criminal offence under the Fraud Act 2006 (see Chapter 14).

Loss for remedial works

Many inexperienced surveyors start off believing that a fully quantified (costed) Scott schedule of dilapidations will set out and identify the level of any loss suffered and that the cost of the works equates to the measure of the damage. In reality, this is rarely so, although the loss associated with undertaking the remedial works will normally form the starting point for any claim for compensatory damages or losses.

Where the remedy being sought includes a claim for damages from the tenant, then the schedule of dilapidations should have each allegation of breach valued and quantified in the most appropriate manner.

Where there is a genuine intention to undertake the works, then the best means to quantify the schedule remedial works items is to have the works competitively tendered by reputable contractors and to use the tender evidence as prima facie evidence of the loss. However, contractors should not be used to price the works as a free quantity-surveying service if there is no genuine intention to undertake works.

Where the landlord may be seeking to mitigate potential dilapidations losses by other means (such as endeavouring to re-let or sell the premises on favourable terms), then carefully prepared surveyors estimates or "pricing" of the schedule items is normally viewed by the courts to provide a reasonable starting measure of any loss subject to subsequent statutory and common law adjustments.

If a surveyor's estimates costing of the schedules are required, then care should be taken as to how individual works figures are appraised. It is common and acceptable practice to use construction works price book information to estimate costs but price book cost data should not be claimed to be definitive and surveyors should be cautious when using such books.

The difficulty with seeking to rely on works price books (such as the BCIS *Dilapidations Price Book* (2nd ed, 2008) is that the published figures are the national mean average gained from a statistical study presented out of context. Unfortunately, most price books fail to provide the full statistical data that would allow the surveyor to identify the relevance or size of the study "population", the date range of the cost data, the standard deviation or the quartile figures, etc. As a consequence, any claim or quantification of loss based on statistical extract data published out of context is often found to be in the range of +/– 30% to the open market local contractors' competitive tender figures that could be obtained.

An alternative and more preferable means to price a dilapidations schedule where works are not being tendered is to employ a local and reputable chartered quantity surveyor to estimate the costs of the works based on their local expert knowledge of recent tender costs.

Regardless of precisely how the schedule of dilapidations remedial works are quantified, the collating of the loss must still have regard to the statute and common law affecting the degree of reasonably recoverable loss (see Chapters 2 and 15).

Consequential losses

Once the loss for the remedial works has been appraised, the surveyors should seek to consider if there have been any further "consequential losses" incurred as a direct consequence of the tenant's breaches.

Consequential losses are those losses which the claimant incurs after the breach has come to the claimant's notice and that result (indirectly) from the tenant's breach. For example, there could be grounds to claim for loss of rent when the property is unavailable for letting and occupation while the tenant's disrepair is being repaired after the end of the lease. While consequential losses are incidental to the main claim, they can form a substantial proportion of the overall loss claimed. In many cases, however, they can also be difficult to prove as having been incurred due to the tenant's breaches.

The decision whether to include an item of alleged consequential loss should be carefully considered, particularly if there is no reasonable evidence directly linking the loss with the tenant's breaches of the lease covenant. Unfortunately, there is also very little case law on the subject. Regrettably, the latest RICS *Dilapidations Guidance Note* (5th ed, 2008), while listing a number of possible consequential loss elements, also fails to provide practical guidance on how consequential losses should be considered.

In order for the surveyor to produce a reasonable claim, the remoteness of the loss must be considered. If the loss flowing from the breach of the lease is too remote then the loss cannot be claimed. To be recoverable, losses must be within the reasonable contemplation of the parties at the time that they entered into the lease. This may seem complicated but there is well-documented case law and commentary on each item of loss.

The landlord surveyor must also bear in mind their client's duty to mitigate their losses, ie to do their best not to increase the amount of the claim against the defendant. If the claimant landlord fails to take all reasonable steps available to minimise their loss, then they will not be able to recover the extra loss that results from their actions.

Where there is reasonable doubt or a lack of immediately available evidence in support of the claim, these can be dealt with by marking the ambiguous but potential claims for loss as "TBC", within any schedule or summary of claim. This reserving of the claim position is particularly relevant if the schedule is to be prepared and served pre-lease end and the landlord simply does not know the full extent of their actual consequential loss that will be incurred at the end of the lease.

Loss of rent

There is little consistency among surveyors in terms of deciding when the grounds for a loss of rent claim exist, and therefore when a claim should be included. This lack of consistency of case law and RICS guidance should make the inclusion of any claim for lost rent a daunting prospect for any landlord surveyor, particularly if they are then to sign a declaration, endorsement and/or statement of truth.

There is a perceived anomaly between the high levels claimed for loss of rent in schedules served on tenants (up to 50% of the total claim) versus the very low sums paid in settlement, but as most cases settle out of court such anomalies rarely get scrutinised or criticised by third parties or professional bodies. When a claim for loss of rent is included in the summary of the claim, the landlord's surveyor often agrees to its omission at the very early stages of negotiation (and this leaves an impression that the claimant has "tried it on", possibly in a fraudulent manner). As the dilapidations community is waking up to issues of fraud and seeking to discipline surveyors when it occurs, it is extremely important for surveyors that they get their claim right. It is also important that surveyors know when to recommend a lawyer's input to clarify this ambiguous area of the dilapidations dispute.

Loss of rates and service charges, etc

If a consequential loss of rent claim can be substantiated, then it naturally flows that a claim for other consequential losses, such as loss of business rates and/or service charges contributions; and/or insurance premium contributions; and/or utility services standing charges, etc should also follow.

For example, if a former tenant's advisors accept that a landlord of a shopping centre has suffered a loss of rent, due to the state of the disrepair of the retail unit, then unless evidence shows otherwise, that landlord should also be able to claim loss of business rates and the loss of service charge payments that would otherwise have been made during the same lost tenancy/rental period by an incoming tenant. In such circumstances, these consequential losses merely seek to compensate the landlord for the costs they have had to pick up as a result of the former tenant's actions and breaches. If making a claim for such losses, care should be taken to make sure it is justifiable and that supporting evidence is obtained and disclosed with the claim.

Loss for professional fees

The landlord will, in most cases, incur professional fees in respect of the preparation and service of the schedule of dilapidations. Whether these are recoverable from the tenant largely depends on whether there is a lease clause which specifically allows recovery. Such a lease clause is common in modern leases but it will need to clearly state which costs can be included in the claim.

The landlord may incur surveyor and legal fees in respect of the assessment of the breaches and the preparation of the schedule. These are often recoverable contractually pursuant to a lease clause and, with reference to *Maud* v *Sandars* [1943] 2 All ER 783, these fees are often recoverable under common law if there is no suitable lease clause.

The landlord may also recover reasonably incurred professional fees on the design and contract administration of the works to remedy the tenant's breaches. Often though, these fees are claimed by the landlord's surveyor at "industry normal" scale fee rates (typically between 8–16% of the net value of the works depending on the scope of works). However, if the contract administration service received by the landlord is less than the full scope of services defined by the "BS1" Scope of Service in the RICS *Appointing a Building Surveyor Guidance Note* (2nd ed, 2000), then any claim for "industry normal" scale fees is open to challenge and reduction.

Any professional fees claimed and the basis for the claims should be clearly set out in the summary of claim. Copies of receipted invoices should also be provided where available.

Loss for "negotiation" fees

The landlord surveyor's fees on the negotiation and settlement of the claim are sometimes itemised on the schedule of dilapidation, with reference to the old percentage-based RICS scale fee. This fee scale was abandoned by the RICS in 1999 at the request of the Office of Fair Trading and is no longer applicable.

Surveyors should apply the rule that negotiation of settlement fees is only recoverable from the tenant if:

- the lease specifically states that they are recoverable
- the dispute is litigated and the courts award costs to the landlord, when the court will have discretion regarding pre-litigation costs. The court would only ever award limited and reasonable costs. The

court would consider the reasonable hourly rate, and apply this according to the time that should have been taken by the landlord surveyor. Such awards of costs for the pre-litigation negotiations of claims that eventually litigate are unusual but possible in the right circumstances. Accurate legal advice will be required.

It should be remembered that "negotiation" fees are fees for professional claim advocacy services where the conduct and appointment fees basis of the negotiators are matters that the courts will take into consideration when determining costs, should the dispute become the subject of formal litigation. If a surveyor accepts an appointment on a scale or performance fee-related basis, then the courts are likely to view such an appointment as posing a conflict of interest. The conflict arises because the performance fee encourages an approach and conduct in negotiations designed to maximise a client's position, rather than to fairly and honestly resolve the dispute by appraising and agreeing the "true measure" of loss and compensation due to the claimant.

If appointed on a scale fee basis, then the surveyor would also be prevented from accepting an appointment as a court-authorised "expert witness", should the matter proceed to litigation. This in turn would cause the client to have to appoint an alternative expert "at the last hurdle" which is arguably a breach of the ethical duty of care owed to the client (see Chapter 7).

Pre-litigation "negotiation" stage advocacy fees should be charged and claimed (where permitted) on a simple "time-charge" basis. Where claims are made for scale or apparently performance-related fee basis, they are likely to be contested (successfully), not only because of the conflict of interest posed but also because the fees claimed may be unreasonably disproportionate to the service actually provided and therefore unreasonable.

Loss for "management time" costs

Many commercial landlords employ in-house building surveyors who negotiate the dilapidations claims as part of their workload. In some, more complex cases, the building surveyor in-house may employ an external building surveyor but still spend substantial time working with the external surveyor to prove the loss.

It can be a source of immense frustration to the landlord's surveyor when the tenant argues that no loss has been suffered in these circumstances simply because no external professional fees have been incurred. This comment can apply equally to the preparation of the schedule or the procurement of building works to remedy the breaches of covenant, ie both areas that are often the subject of valid claims by landlords against tenants.

The answer to the problem is that the in-house surveyor should keep detailed time records in the same way that an external surveyor would record their involvement. Instead of using these records to justify the fees to a client, these records are used to justify time engaged in dealing with the claim to the opposing surveyor and potentially the courts.

Where the opportunity for a contemporaneous record has been missed, retrospective assessments are considered to be a valid but less preferable method of calculation. Such an assessment can only be an approximation of the hours spent and may over-estimate (or under-estimate) the actual time which would have been recorded at the time.

The courts will err on the side of caution when considering any claim based on such a forensic analysis and will readily discount a landlord's claim put in this form, ie all the more reason for producing very detailed time sheets right from the outset. The appropriate approach to the question of recovery of wasted landlord management time was neatly summarised in *R & V Verischerung AG* v *Risk Insurance & Reinsurance Solutions* [2005] EWHC 1253, which held that:

> as a matter of principle, such head of loss (i.e. the costs of wasted staff time spent on the investigation and/or mitigation of the tort) is recoverable, notwithstanding that no additional expenditure "loss", or loss of revenue or profit can be shown. However, this is subject to the proviso that it has to be demonstrated with sufficient certainty that the wasted time was indeed spent on investigating and/or mitigating the relevant tort; i.e. that the expenditure was directly attributable to the tort ... that to be able to recover, one has to show some significant disruption to the business; in other words that staff have been significantly diverted form their usual activities. Otherwise the alleged wasted expenditure on wages cannot be said to be "directly attributable" to the tort.

The above judgment principle, while relating to a claim in tort, was subsequently applied to a construction works related contractual damages claim (as a consequential loss) in the case of *Bridge UK.Com Ltd (t/a Bridge Communications)* v *Abbey Pynford plc* [2007] EWHC 728 (TCC).

The opportunity cost in respect of management time should always be considered by the landlord. If reasonably incurred and clearly documented, it can form a valid head of claim in a lease end dilapidations dispute.

Loss for waste

While it is rare to see a claim present that includes a claim for losses for waste, it remains possible to do so under the ancient Statute of Marlborough (1267) and the law of waste (see Chapters 5 and 7).

Loss of amenity

Another rarely encountered but possible claim for loss concerns the damages associated with a loss of an "amenity". For example, a large roof light might have been removed during a tenant's over-roofing "repair" programme, where the new roof is reasonable but where natural daylight and fresh air to the centre of a "deep" floor plate within a large space may have been lost. This in turn may make the internal accommodation/space less desirable to prospective tenants and (if left) could also in time result in loss of a right of light. The loss of the roof light could therefore conceivably reduce the amenity of the space.

Perhaps the most well known loss of amenity case was *Ruxley Electronics and Construction Ltd* v *Forsyth* [1995] EGCS 117, which concerned the "loss" of six inches of depth to a swimming pool during construction. The claim went all the way to the House of Lords and all the contractual damages elements of the claims failed. In the end, however, the claimant was awarded token damages for the loss of amenity associated with having received a pool reduced in depth by six inches (which was dwarfed by the costs the claimant suffered in taking the dispute to the House of Lords).

Surveyors should be mindful but wary of issues such as loss of amenity as the levels of losses awarded if successful are often subsumed by the costs in bringing the claim in the first place. Where there is potential for a loss of amenity claim, surveyors should seek client instructions and encourage a pragmatic approach where possible.

Loss for Value Added Tax (VAT)

The recovery of damages in lieu of VAT in dilapidations claims are a widely misunderstood and misapplied element of many landlord dilapidation claims, yet the logic behind whether damages are recoverable is relatively straightforward to establish.

The ability to claim for loss of VAT incurred in relation to dilapidations related works costs, consequential losses and fees, etc will depend on whether the claimant landlord is able to recover any VAT incurred from HM Revenue and Customs (HMRC). If they can recover VAT from HMRC, then they will not have suffered the VAT as a loss. However, whether they can recover the VAT will depend on what surveyors commonly refer to as the landlord's "VAT elected" status.

Elected for VAT principles

Section 1.3 of HMRC Notice 742a *Opting to Tax Land and Buildings* (June 2008) also states:

> Supplies of land and buildings, such as freehold sales, leasing or renting, are normally exempt from VAT. This means that no VAT is payable, but the person making the supply cannot normally recover any of the VAT incurred on their own expenses.
>
> However, you can opt to tax land. For the purposes of VAT, the term 'land' includes any buildings or structures permanently affixed to it. You do not need to own the land in order to opt to tax. Once you have opted to tax all the supplies you make of your interest in the land or buildings will normally be standard-rated. And you will normally be able to recover any VAT you incur in making those supplies.

When surveyors refer to the claimant's "VAT elected status" they are actually referring to whether or not the landlord claimant has "opted" (elected) to tax the specific lease demised land and property interest (see above). If they have opted to tax then they will be able to recover any VAT on land and property related services at the standard rate, currently 17.5%. However recovery rates may vary as the individual lease demised property may be part of a landlord's large portfolio, where the other properties may have a mix of elected/exempt statuses and HMRC may have agreed a general portfolio average recovery rate that reflects the cumulative portfolio elected status. Where the property is "elected", the act of recovery may be possible by offsetting VAT sums paid but to be recovered against other VAT sums due to HMRC in the landlord's VAT accounts.

In addition, section 10.10 of the HMRC Notice 742 *Land and Property* (March 2002), states that:

> A dilapidation payment represents a claim for damages by the landlord against the tenant's 'want of repair'. The payment involved is not the consideration for a supply for VAT purposes and is outside the scope of VAT.

In other words, the act of receiving a dilapidations related damages payment from a tenant is not a service that incurs VAT in its own right, but is merely a receipt of compensation. However, this is not the same as saying that non-recoverable VAT losses cannot form part of the compensation payment.

When can the landlord recover or offset VAT?

The landlord's schedule should be a fair and accurate statement of the landlord's loss such that the tenant can understand the extent of the full liability if the tenant fails to complete the required works as soon as is reasonably possible. This is a requirement of the Property Litigation Association's *Dilapidations Protocol* and essential information to settle any dispute.

If the landlord (or superior landlord) has not elected/opted to tax the particular lease demised property in question, then they are legitimately entitled to seek recovery from the tenant of VAT costs incurred on the dilapidations claim service elements. These elements may include remedial construction works services, professional fees, consequential service losses, etc.

If however, the landlord can recover the VAT element from HMRC, then the landlord cannot make a claim on the tenant for the VAT expense. If the landlord were to do so then the landlord would effectively achieve a "double claim" (recover the same VAT from both the tenant and HMRC) and this would go against the common law principles of loss.

If the landlord is not doing the works, eg the landlord is selling or just passing the dilapidations payment to an incoming tenant as a rent-free period, then the damages received would be classed as a capital receipt and tax would be paid to HMRC.

Negligent or fraudulent claims for VAT

Surveyors must understand that the landlord claimant's VAT status will be a question of *fact* in every case. Any claims for VAT will therefore be statements of fact and it is therefore essential due diligence for surveyors to check and establish the VAT elected status and any recovery rate *before* a claim is made. Often, the VAT elected status can be quickly established by checking if VAT is charged on the tenant's quarterly rent payments (and if so at what rate). If charged, it is a good indicator that the property is elected for VAT and so VAT should not feature in the claim.

The misapplication of this element of the claim can create an embarrassing disparity between claim and true loss and eventual settlement (and may even delay settlement and the receipt of damages by the landlord). If the surveyor includes illegitimate claims for VAT either on a knowingly false, reckless or uncaring basis, they will in effect make an exaggerated claim (by up to 17.5%) which could be considered to be "fraudulent misrepresentation" or, depending on the circumstances, a criminal act of fraud under the Fraud Act 2006.

If litigated, the courts would immediately reject any erroneous claim for VAT and the claimant and their surveyor's credibility and competence would be damaged to the detriment of the remainder of any genuine claim.

Considering diminution

Legal restrictions on loss

Section 18(1) statutory diminution restriction

The "first limb" of section 18(1) of the Landlord and Tenant Act 1927 states:

> Damages where a breach of covenant or agreement to keep or put premises in repair during the currency of the lease, or to leave or put premises in repair at the termination of a lease, where such covenant or agreement is expressed or implied, and whether or specific, shall in no case exceed the amount (if any) by which the value of the reversion (whether immediate or not) in the premises is diminished owing to the breach of such a covenant or agreement as aforesaid.

Common law diminution restrictions

Over the years, the interpretation and application of the first limb of section 18(1) has been widened by common law and it is currently held to apply to all claims of repair, structural, or otherwise; and also to "decorative repair" as recently confirmed by the case of *Latimer* v *Carney* [2006] 3 EGLR 13. Common law has also sought to apply wider general restrictions on the level of recoverable damages similar to those imposed by section 18(1), particularly in claims where it is alleged that the "cost of works" should be treated as the measure of the damages. Famously, Lord Lloyd of Bewick in *Ruxley Electronics and Construction Ltd* v *Forsyth* [1995] EGCS 117 stated:

> **first**, the cost of reinstatement is not the appropriate measure of damages if the expenditure would be out of all proportion to the benefit to be obtained, and, **secondly**, the appropriate measure of damages in such a case is the difference in value, even though it would result in a nominal award.

Diminution valuations

General approach to diminution valuations

Since no two properties are identical and diminution valuations are rarely straightforward, it is not surprising that surveyors, lawyers and the courts have sought a standardised process that can be adopted in the majority of lease claims. In recent years, dilapidations surveyors have often adopted the diminution valuation approach appended to *Shortlands Investments Ltd* v *Cargill plc* [1995] 1 EGLR 51.

The "Shortlands" approach of appraising the value of the diminution of the reversionary interest relies upon the preparation of two valuations. It is a clear process and dilapidations surveyors and lawyers should be familiar with the ruling. This method of appraisal commonly requires the following.

- Valuation A: an initial valuation to assess the open market value of the property at lease end with no breaches of the lease.
- Valuation B: a second valuation that seeks to appraise the value of the property in the open market, with allowances being made for a tenant's breaches that would survive a hypothetical purchaser's future intentions for the property.

The minimum difference between valuation A and valuation B is often said to be the "diminution in value" of the reversion attributable to the tenant's breaches. However, surveyors should be aware that the courts do not always support this approach.

In theory, the "Shortlands" approach seems straightforward but in reality, valuation issues on commercial property are often difficult to assess. Surveyors seeking to adopt the "Shortlands" approach need to understand the risks involved in doing so. For example, the courts have held that it can result in calculating a hypothetical loss without first establishing if the loss actually exists.

Also, if care is not taken, then an incomplete appraisal can result which distorts the measure of loss rendering the calculated diminution in value of little or no merit. Valuers should ensure that they consider not only the normal "Shortlands" valuation factors, but that they include within the valuation an appraisal of vendor transaction costs, such as capital gains payments that would be associated with a sale of the title or other tax issues.

Specific to its circumstances

Where there are reasonable and proportionate grounds for preparing formal diminution valuations, due care and diligence must be taken to determine the damage to the vendor's reversion. The apparent lack of detailed comprehension of the "Shortlands" approach frequently leads to dilapidations surveyors, valuers and lawyers adopting overly formulaic and inflexible or irrelevant valuation positions on some claims.

In some cases, inadequately considered diminution valuations may lead to protracted dispute resolution negotiations. Poor diminution appraisals may also lead to the later collapse of the damages claim or defence positions once litigation proceedings have been formally commenced with significant cost implications for the losing party.

Formulaic spreadsheet-based valuation approaches can be a suitable starting point and provide a useful partial checklist of valuation considerations. However, dilapidations surveyors and valuers must remember that each valuation has to be unbiased, carefully considered, and diligently prepared if it is to be relevant to the specific property and circumstances.

Is a formal diminution valuation required?

A significant degree of emphasis is often placed upon the need for the preparation of a formal diminution valuation at lease end. The issue of formal valuation must be kept in context. For example, the judge in *Latimer* v *Carney* [2006] 3 EGLR 13 restated the common law position that:

> this court (the Court of Appeal) in *Jones* v *Herxheimer* did not consider that it was necessary for there to be formal valuations done by experts in

every case, and in addition expressly envisaged that the judge might not accept the evidence of an expert valuer and thus by implication reach a valuation on some other basis.

In circumstances where the landlord does not intend to undertake remedial works and when commencing the preparation of diminution valuations, surveyors must first consider whether there is evidence of any damage to the reversionary value that warrants the production of a formal set of valuations.

In considering this aspect further, the surveyor(s) should seek to take into consideration the landlord's future marketing/letting intentions for the property. Where possible to do so, the surveyors should also briefly and informally discuss the property with a local and experienced valuer or agent to ascertain the outline expectations and sale or re-letting terms of a reasonable minded purchaser/tenant for the property in the local market.

Commissioning diminution valuations

Where enquiries establish that it is probable that damages to the reversion are likely to have been suffered and where an appraisal of diminution is beyond the surveyor's personal competency, then consideration should be given to obtaining a formal diminution valuation from a suitably experienced local expert valuer.

Where local expert diminution valuations are to be commissioned, in order to mitigate costs, both landlord and tenant parties should wherever possible seek to agree the appointment of a single joint expert valuer to provide a single valuation.

As valuation is not an exact science but is more an "art", a single joint expert valuation should help avoid the commonplace "battle of the experts" that typically follows when two separate partisan valuers produce conflicting diminution valuations. Typically, battle-of-expert valuations merely serve to cause a secondary dilapidations dispute between the parties, thereby reducing the likelihood (or at least the perception) of a fair and reasonable settlement being achieved out of court.

Any valuation obtained should be openly disclosed to the other party to the dispute as soon as available and will be disclosable if the dispute proceeds to formal litigation and court proceedings.

Each valuation obtained should be a complete and stand-alone document that contains not only the "Shortlands" valuations A and B; but should also contain an appraisal of the local market, complete with comparable and relevant sale/letting transaction evidence. The valuation report should also make clear the valuer's assumptions and opinions on the "reasonable minded" local purchaser or tenant's expectations and future occupancy or use for the property.

Where the valuer considers it likely that a prospective purchaser would consider alternative uses for the property; and where it is the landlord's stated intention to sell the property, the valuers may need to provide further additional valuations and appraisals for the property for each potential and reasonably likely/possible alternative changes of use.

Applying diminution "caps"

Once it has been established that there has been a diminution in the reversionary interests attributable to the tenant's breaches and wants of compliance with their lease terms; and once the level of the diminution has been appraised and calculated, it remains for the diminution cap to be properly applied.

It is common to find that any calculated diminution "cap" is routinely and often erroneously applied to the entire schedule of dilapidations claim under the pretext of being a cap imposed by section 18(1) of the Landlord and Tenant Act 1927, *regardless* of whether or not the entire claim concerns only breaches of repair and decorative repair covenants (see Chapter 15).

More care will need to be taken where the schedule of dilapidations includes items of alleged breaches of clauses other then repair and decorative repair clauses, such as breaches of alteration or reinstatement clauses or wants of statutory compliance.

Where other non-repair or non-decorative repair breaches are being alleged, then the costs for the repair and decorative repair items, together with a pro-rata share of any consequential or other loss being claimed, will need to be separated out from the rest of the claim. The statutory section 18(1) diminution "cap" must then be "applied" only to the sub total of the repair and decorative elements of the claim.

Once a statutory cap has been applied, then careful consideration will need to be given in applying any common law cap in accordance with the diminution valuation principles set out above. Where the cost of the works (whether undertaken or not) is either unreasonable or disproportionate to undertake by comparison to any diminution in value that may be suffered if the works are not undertaken, then the entire claim will find itself capped by the common law principles of loss (see Chapter 2).

Summarising the claim

Scheduling the loss

Once the various potential losses and restrictions on losses have been appraised, the surveyor will need to compile and summarise both their own and third party claim or loss evidence into a succinct and easy to follow summary of claim document. A typical dilapidations summary of claim may be structured as follows:

Example summary of claim

Schedule of dilapidation remedial works summary			
Wants of repair schedule items:		£#	
Wants of decoration schedule items:		£#	
Wants of alteration reinstatement schedule items:		£#	
Wants of compliance schedule items:		£#	
Other miscellaneous schedule items:		£#	
Schedule remedial works sub total			**£#**
Note 1: Proportion of repair and decorative repair items to the schedule sub total	##%		
Schedule remedial works contractor costs			
Contractors "preliminaries" at ## %	£#		
Contractors "overheads and profit" at ## %	£#		
Less TPI adjustment for contract losses from tender date back to lease determination date	–£#		
Remedial works sub-total		**£#**	
Non-recoverable VAT		£#	
Remedial Works contract sum			**£#**
Remedial works contract professional fees			
Contract administration — RICS BS1 services stages A and B	£#		
stages C to H at ## % of the net value of the works contract subtotal	£#		
"Additional services" charges incurred	£#		
Contract administrator sub total		**£#**	
CDM planning co-ordinator at ##% of the net value of the works contract sum		£#	
M&E consultant/designer professional fees and costs		£#	

Item			
Structural engineer consultant/designer professional fees and costs		£#	
Quantity surveyor consultant professional fees and costs		£#	
Other consultant professional fees and costs		£#	
Building control fees (based on £##k notifiable works)		£#	
Works project consultant fees sub-total		**£#**	
Non-recoverable VAT on works related fees and costs		£#	
Remedial works professional fees sub-total			**£#**
Claim preparation and service fees			
Landlord's surveyor dilapidations survey and schedule of dilapidations production		£#	
Landlord's third party survey/inspection and reporting costs		£#	
Landlord's valuer's diminution valuation fees and costs		£#	
Landlord's solicitor claim service		£#	
Non-recoverable VAT		£#	
Sub Total			**£#**
Consequential losses appraisal			
Vacancy losses (for a duration of ## weeks):			
Loss of rent at £## pa	£#		
Loss of rates at £## pa	£#		
Loss of insurance premium contributions at £## pa	£#		
Loss of services standing charges at £## pa	£#		
Loss of building management charge contribution at £## pa	£#		
Vacancy losses sub total		**£#**	
Loss of amenity (of ##)		£#	
Loss for wasted "management time"		£#	
Loss for waste		£#	
Other losses incurred (... ...)		£#	
Loss of interest at #% APR		£#	
Non-recoverable VAT		£#	
Sub total			**£#**
Gross pre-cap claim			**£#**
Statutory claim cap appraisal			
Proportion of gross pre-cap claim for repair and decorative repair items — see note 1 above	%		
Repair and decorative repair portion of the gross pre-cap claim (for s 18(1) cap application)	£#		
Calculated diminution loss	£#		

First limb s 18(1) capped repair and decorative repair claim (*note: lesser of the two lines above*)	£#	
The remainder of the gross pre-cap claim	£#	
Section 18(1) applied claim sub-total		**£#**
Common law cap		
Calculated diminution loss (same as above)		£#
Gross reasonable and proportionate damages claim		**£#**

Note 2: subjective appraisal but possibly the lesser of s 18(1) applied claim sub-total; or the calculated diminution loss.

Where elements of the above example summary of claim are irrelevant or not applicable, then they should be omitted from the statement to avoid confusion. Care should also be taken to ensure that any spreadsheet formulas used within the table calculate correctly.

The damages claim

The summary of claim document should also clearly set out the following information:

- The landlord and tenant parties to the dispute and claim including their respective names and addresses.
- The summary of claim and schedule of dilapidations documents (see above).
- The gross damages sum being claimed from the tenant, together with a breakdown of how the damages have been appraised (such as listing the summary of claim (see above).
- Any documents, evidence or valuations to be relied upon by the landlord, including, where available, copies of any receipted invoices or other evidence of costs and losses.
- A date by which the tenant should reasonably be required to respond to the claim.
- The Civil Procedure Rules Practice Direction — protocols section 4 default protocol required statements (see Chapter 3).

11 Serving the Schedule

Schedule service considerations

Service of documents

The schedule should be served by the landlord's lawyer wherever possible. The point at which the schedule is issued by the surveyor to the lawyer should be the point that a second opinion is sought by the surveyor in terms of the interpretation of relevant lease clauses and the inclusion of all lease and licence obligations.

The primacy of the landlord's first schedule cannot be overstated. The form and content of the schedule is crucial. In the event that the matter results in court proceedings, the trial judge may well read the initial claim to assess the landlord's lease end intention. The original version of schedule claims and responses will also be of material relevance when it comes to awards for costs.

At the stage of service the landlord's lawyer should:

1. Scrutinise the claim on an elemental basis to ensure that all areas of the demise are covered in a clear and logical fashion.
2. Ratify the technical and legal logic of the claim. For example, where supersession or improvement applies, has this been dealt with by the surveyor in a clear and well reasoned fashion, or does the quantum of the claim reflect the likely or actual construction costs?
3. Assess the claim for consequential losses. These must be accurately stated, with evidence to back up elements claimed. Consequential losses can equate to 50% of the claim (loss of rent, fees, rates, damages in lieu of VAT) and it is vital that the claim is correct.

4. Confirm that the surveyor has checked the landlord's intentions for the demised premises and that they will actually be suffering the losses claimed. Has the surveyor spoken in detail to the landlord client and has the crucial issue of landlord intention been factored in the claim?

The lawyer also has an ethical duty to protect the client from ill-considered actions. The lawyer can protect their client best if they interrogate the surveyor to find the weaknesses in the claim at the early stage. While the time taken to complete the above will often be greater than the lawyer's time charge for simply serving the schedule in accordance with the lease provisions, it is often time well spent. Even on the smaller claims the use of a second opinion on the legal logic of a claim can save time, expense, and costs further down the line.

In most commercial property leases there is a clause allowing for the recovery of a landlord solicitor's fees incurred in the service of the schedule. The lawyer will generally be able to recover these fees and the lease should always be checked. In addition, the default position of involving a lawyer to serve the notices and schedules should be adopted, unless strong reasons exist to the contrary.

In respect of reinstatement notices and interim schedules of dilapidations, there are yet more compelling reasons for the lawyer to serve these documents. For example, it may be vital that time critical notices are served by the correct date, in a format that satisfies the lease clause.

The lawyer and surveyor working for the landlord must always remember that the service of the dilapidations schedule is the initiation of a legal damages claim and for this reason alone a lawyer should be responsible for this stage of the claim.

Taking the above measures will start the claim on the right footing which can only help create a well ordered claim, which in turn facilitates settlement.

Timing of the service of documents

If a surveyor adopts the approach suggested by the Property Litigation Association's (PLA) *Dilapidations Protocol*, then the schedule should be served by the landlord "within 56 days of the end of the lease". This is often unworkable and unrealistic for several reasons:

1. It is overly prescriptive, terminal schedules can be served at any time during the last three years of the lease term (by virtue of section 146 of the Law of Property Act 1925).
2. Tactically, the landlord may want to serve the schedule early to encourage the tenant to do the works. Conversely, the landlord may not know what works they want to be done to the premises until lease end and, therefore, the service of a heavily caveated theoretical schedule will be of little use. If the works are substantial, the time limit of 56 days post lease end may not be achievable.
3. The landlord may not trust their tenant to complete the works to a suitable standard and may reasonably adopt a reasonable "reactive only" stance to the issue of what works need to be done right up to the end of the lease.
4. Leases commonly determine on the rent quarter dates which typically coincide with the common holiday periods/seasons in England and Wales. This can make it difficult to gain timely quotes, instructions or professional resources and services. It often means that there are longer mobilisation periods while people are unavailable due to holiday commitments. If all dilapidations issues are to be carefully and skilfully considered and scheduled, then the more complex claims may not be possible to prepare within the 56-day timescale.
5. If tenders for works are required then the Joint Contracts Tribunal's (JCT) good practice guides for single stage competitive tenders suggests that tendering contractors are provided with four weeks to submit tenders. If works are being undertaken, the remedial works duration in many claims may exceed 56 days. So where works are being undertaken, it is more likely than not that the claims cannot be made within 56 days.
6. Because schedules can theoretically be served at any stage from three years before the lease, up to 6 or 12 years after (depending on the limitation period), the exact date of service is a matter of value judgment for the landlord.

While the PLA's suggested timescale is in most cases unachievable, the claim should be served at the earliest opportunity that case specific circumstances permit. Assuming all other factors remain equal, the landlord should try to serve the claim at or close to lease end, as this will be compelling evidence of the landlord's intention at the crucial lease end date.

Surveyor declarations

It is becoming good practice that the schedule or claim (or both) include suitable surveyor declarations similar to those required to be made by "expert witnesses". The surveyor's opinions within the schedule and claim documents will be their "expert" opinion as the surveyor's document is normally prepared in contemplation of litigation and will be an admissible document (see Chapter 4). When making a declaration, the surveyors should also disclose potential conflicts of interest in keeping with principles described in the case of *Toth* v *Jarman* [2006] EWCA Civ 1028.

The surveyor may wish to consider using the following sample good practice set of declarations in pre-litigations schedule and claim documents:

I DECLARE THAT:

1. I have endeavoured to include in this schedule those matters, of which I have knowledge or of which I have been made aware, that might adversely affect the validity of my opinion. I have clearly stated any qualifications to my opinion.
2. I have shown the sources of all information I have used.
3. I have not, without forming an independent view, included or excluded anything which has been suggested to me by others, including my client or their lawyers.
4. I will notify my client and confirm in writing if for any reason this schedule requires any correction or qualification.
5. I understand that:
   (a) This schedule, subject to any corrections before swearing as to its correctness, may be used as evidence in future litigation if the dispute is not resolved amicably between the parties.
   (b) I understand that if this matter becomes subject to litigation and a court hearing, that I may be cross-examined on this schedule by a cross-examiner assisted by an expert;
   (c) I understand that I am likely to be the subject of criticism by a judge if this matter is litigated and if the Court concludes that I have not taken reasonable care in trying to meet the standards set out above.
6. I confirm that:
   (a) I have no conflict of interest of any kind, other than any which I have disclosed in this schedule.

(b) I do not consider that any interest I have disclosed affects my suitability on any issue on which I have acted or that may be called to give evidence upon.
(c) I will advise my client if, between the date of this report and any subsequent trial, there is any change in circumstances which affects my declarations to 6a or 6b above.

7. I confirm that I have not entered into any arrangement where the amount of payment of my fees is in any way dependent on the outcome of the dispute.

Statements of truth

The PLA *Dilapidations Protocol* has sought to introduce a surveyors "endorsement" to accompany schedule of dilapidations documents. In the 2006 version of the PLA Protocol, it required a surveyor's declaration concerning the quantum of the claimant's loss.

This endorsement was much criticised and widely ignored by surveyors. Within the May 2008 version (version 3), the PLA endorsement produced a declaration to be signed by the surveyor that they have exercised due care and skill in providing their professional services (although the purpose and benefit of this latest endorsement is unclear).

Despite the recent revisions to the PLA Protocol and the Royal Institution of Chartered Surveyors (RICS) *Dilapidations Guidance Note* (May and June 2008 respectively), the proposed surveyor endorsements remain contentious and appear to add little in establishing the truthfulness of a surveyor's opinion of the claim from the outset.

An alternative statement of truth is contained in the practice direction on the Civil Procedure Rules (CPR) Part 35 as follows:

> I confirm that insofar as the facts stated in my report are within my own knowledge I have made clear which they are and I believe them to be true, and that the opinions I have expressed represent my true and complete professional opinion.

Due to the lack of a CPR "approved" dilapidations specific pre-action protocol, with its own particular version of the statement of truth, it is open to surveyors to further develop and adopt suitable best practice variations of the wording for any statement to be signed at the base of the schedule. In the meantime, however, it is unlikely that a surveyor could be criticised in most dilapidations circumstances for making a CPR compliant statement of truth using the CPR court-approved "expert" version quoted above.

Part 36 offers

Dilapidations can be an uncomfortable mixture of negotiation and expert evidence. While the RICS and the PLA work to clarify the uneasy hybrid that is "dilapidations dispute resolution", surveyors and lawyers working with their clients must try to agree settlements. Perhaps the best way to push for settlements is to issue a Part 36 offer at or near to the time that the schedule is served (also see Chapter 13).

The exact timing of an offer is often a matter of personal preference. Either way, the existence of such an offer at the inception of a dispute has several advantages. Most importantly, the offeror receives protection on costs if correctly pitched in the event that the dispute continues to court. The secondary impact of an early offer is to give the recipient of the schedule the clear impression that there is a deal to be done. It can transform a drawn out dispute into a reasonable and straightforward claim and settlement, for the benefit of all parties.

The surveyor or lawyer should always consider issuing versions of a Part 36 offers as follows:

1. A financial offer for the settlement of the entire quantum of the claim.
2. A financial offer for individual (preferably substantial) elements of the claim.
3. An offer in respect of a substantial liability (costs) issue for an element of the claim.
4. An offset proposal for any counter claim to be made for costs incurred in defending an unreasonable or unjust claim.
5. A combination of the offers.

The wording of the Part 36 offer should be checked with the landlord's lawyer at the time of service, with the offer ideally being issued via the lawyer with the main schedule, by way of a "without prejudice" side letter.

12 The Tenant's Counter Schedule

Introduction

The tenant's surveyor owes their client a professional and ethical duty of care, but must not allow their desire to minimise the claim override their obligations under the Civil Procedure Rules (CPR) 1998 and their role in seeking a fair and genuine settlement. Balanced against this, the tenant's surveyor should seek to test, explore and professionally challenge the landlord's claim before forming their firm opinion of the claim and reaching a settlement.

Tenant surveyor appointment

In any claim, both landlord and tenant surveyors will be engaged on a common matter that is dependent on one particular property or estate and that sits within the same lease and legal framework. Consequently, when commencing their services, the initial appointment of a tenant's surveyor will be virtually identical to those previously described for the landlord's surveyor within Chapter 7.

So the tenant's surveyor can then appraise the landlord's claim independently and should consider issues such as:

- potential conflicts of interest
- public indemnity insurance cover
- terms of appointment.

Due diligence checks

The tenant's surveyor should carry out the following checks prior to commencing the preparation of their formal response:

- Identifying the type of the tenancy.
- Obtaining and checking lease.
- Checking the validity of a lease.
- Obtaining and checking other tenancy documents.
- Lease privity and status checks.
- Checking Her Majesty's Land Registry (HMLR) registered interests.
- Checking landlord and tenant status.
- Checking who's paying the rent.

Reviewing the tenant's obligations

The tenant appointed surveyor should seek out and review all documentation and material facts relevant to the dilapidations claim. These will include:

- Determining the extent of the demise.
- Repairing and redecoration covenants.
- Alteration and addition covenants.
- Clarifying or seeking to establish a detailed description of the demise at the start of the lease (layout, fixtures, fittings, etc).
- Appraising reinstatement obligations.
- Legal compliance covenants.
- Other common tenant's covenants.
- Schedules of condition.
- "Tenant-like" user obligations.
- Reasonable market expectations.
- Other pre-survey complicating issues.

Further guidance on the above issues can be found in Chapter 7.

Checking the landlord's works intentions

Intention considerations

The importance attached to the landlord's intentions for the property or estate at the material date for the claim cannot be underestimated.

The reasonableness and proportionality of the landlord's proposed remedy and/or the level of damages that the landlord may claim, will be significantly dependent upon the consequences of their intentions. It is therefore of primary importance that the tenant's surveyor seeks to understand and critically examine the landlord's intentions.

In general, the courts will take into consideration the "fixity" of intention at the material date of the claim. Therefore, if the landlord has not formed a fixed intention at the material date to, say, demolish the premises until after the termination date, they can still recover damages even though they may, at a later date, consider an alternative scheme and subsequently demolish the property.

Conversely, the landlord may be prevented from recovering damages if there is a known scheme to redevelop or demolish a property at some stage after the end of the lease, which in some cases has even been as remote as five years after the material date in a claim.

In some cases, where there has been doubt as to the genuineness of a landlord's alleged intentions for a property, the courts have allowed examination of directors' board meeting minutes and company files for a considerable period prior to the material date of the claim where, on occasion, evidence has been found as to the true (and often contradictory) landlord intentions.

Checking the landlord's intentions

The landlord's claim documents should be checked to see if they contain a clear statement of their intentions for the property. This may possibly be found in the claim cover letter or alternatively may be contained within the body of the schedule of dilapidations. Increasingly however, it is becoming good practice for landlord surveyors to include a clear declaration of a landlord's intentions within the preambles section of the schedule of dilapidations document.

Where a landlord's statement of intent has been made, it should be checked to ascertain whether or not it is an absolute statement of a committed intention; or whether it is a less committed intention where perhaps the landlord has declared that they intend to follow one option but reserve their rights to pursue an alternative course in order to try and mitigate a claim but reserve the right to follow an alternative course of action should it become necessary.

The tenant's surveyor should also check whether there was any previous statement of intentions by the landlord that may either support or contradict the claim stated intentions. For example, has the landlord submitted any applications or notices to the local authority for statutory consents or approvals that may indicate an intention to redevelop or substantially alter, extend or refurbish the property in question? Also, has the landlord made any alternative statements of intention in a section 25 Landlord and Tenant Act 1954 notice shortly prior to the end of the term?

Further intention checks can include checking if the landlord is actively marketing the property for sale or re-letting. If so, requests should be made to obtain disclosure of the full marketing details available to prospective purchasers or tenants and a copy of the marketing agent's marketing report to their client (the landlord).

It is also prudent to make a formal request to the landlord and their solicitor that they openly disclose details of any sale or letting negotiations that are ongoing; and that they keep the tenant's surveyor openly informed of developments in this regard during the claim. Where sale or re-letting negotiations are ongoing or commence, the landlord should be asked to openly disclose copies of any agreements reached or written heads of terms.

If the landlord is alleging that they intend to undertake the remedial works identified within the schedule of dilapidations, they should be informed that they do so "at risk". Where works are planned, the landlord should be asked to disclose full copies of tender documents, including professionals' appointment letters, tender specification of works documents, invitations to tender, tender returns, post- tend clarification correspondence, tender reports and contractor appointment/works contracts.

Challenging stated intentions

Where examination of the landlord's stated intentions identify discrepancies or possible conflicting intentions; or where the surveyor considers the intentions unrealistic or dubious for whatever reason; then the tenant's surveyor should seek to openly raise their concerns over the landlord's claim stated intentions at the earliest opportunity.

If raising concerns over the stated intentions, it is prudent to remind the landlord and their advisers that their statement of intention is a materially relevant issue and that any statement they make must be factually true. It is not unreasonable to seek to remind the landlord and their advisers that any misrepresentations made either recklessly or knowingly may have serious consequences for the claim and any settlement reached based on a misrepresentation.

If raising concerns over stated intentions, the landlord should be invited to review their own claim documents again; and to once more openly reaffirm or revise their intentions by providing a suitable "statement of intentions".

Once the landlord reaffirms their intentions, then the parties should seek to attend to the claim proceedings on the basis of the openly stated intentions. Should evidence then come to light at a later stage that demonstrates that the statements were false; the tenant will be able to seek the overturning of any settlement reached on the basis of false statements. In such circumstances, the landlord and/or their advisors who recklessly or knowingly made the false statements could find themselves subject to investigation by the proper authorities and may even find that they have committed a criminal offence under the Fraud Act 2006 and other statutes. The issue of misrepresentations and fraudulent claims is further covered in Chapter 14 of this book.

Checking the landlord's required remedy

One of the first and most important tasks for a tenant's surveyor to undertake is to establish exactly what remedy the landlord is seeking to resolve the claim. For example, has the claim been served in such a manner so that the landlord is seeking possible forfeiture of the lease and/or damages; or is the landlord intending to enter onto the tenant's property, undertake "self-help" remedial works and recover the costs from the tenant as a debt?

The tenant's surveyor should ensure that they obtain a copy of the landlord's cover notice or letter that accompanied the schedule and statement of claim documents and review it to establish the precise nature of the remedy being sought (see Chapter 5 for possible remedies).

Section 146 notice action

If the landlord has served a notice under section 146(1) of the Law of Property Act 1925 seeking forfeiture; or notifying an intention to "re-enter" and determinate and/or damages, then it may be possible for the tenant to challenge the validity of the notice or seek the benefit of the Leasehold Property (Repairs) Act 1938 and gain "relief" from the landlord's actions.

The content of the notice

Any section 146(1) notice should satisfy the requirement of the Act in that it should state/refer to:

1. The particular breach(es) complained of.
2. If the breach is capable of remedy, the remedial action or works action required of the tenant.
3. If compensation (damages) is being sought, what sums are being sought.
4. The reasonable period in which the tenant has to comply with the notice.
5. Where the tenant may be able to claim relief under the LP(R)A 1938, then the notice should inform the tenant of their rights under the 1938 Act to claim the benefit of that Act and relief from the landlord's actions within 28 days of the date of the notice.

Points 1 and 2 are normally adequately covered by the text of a landlord's schedule of dilapidations but the schedule should be clear of the precise lease clauses allegedly breached. If the notice lacks any details on damages, the damages point 3, then the landlord may have "waived" their right to claim damages under that notice even if they then successfully gain forfeiture or re-enter and determine the lease.

Where the lease contains a "self-help" clause, then the "reasonable" period stated in point 4 for the tenant to comply with the section 146 notice should also be checked against the reasonable or minimum period for tenant compliance with notices to repair, likely to be expressly stated within the self-help clause. If the section 146 notice period is for a shorter duration and where there is no extraneous circumstances that would justify the shortened reasonable period, then the notice period could be challenged by the tenant.

If any discrepancies are found with the notice, then the tenant's legal advisers should advice their tenant on their defence options and, where necessary, challenge the validity of the notice at the earliest opportunity.

The ability to claim general "relief"

If the section 146 notice has been validly served and is of sufficient and appropriate content, then the tenant's surveyor should consider if relief is available under the Leasehold Property (Repairs) Act 1938.

There is a useful flowchart overview of the rights for relief under the 1938 Act in Chapter 5.

Time is of the essence for the tenant to claim relief under the 1938 Act and every care should be taken to ensure that the 28-day deadline for claiming relief is not missed (although relief from forfeiture may still subsequently be gained if the tenant then commits to suitable undertakings so that the breaches for which forfeiture is being sought will be addressed).

Where the tenant can't claim the benefit of the Leasehold Property (Repairs) Act 1938 or gain relief from forfeiture, then the tenant should give serious consideration to commencing at least a partial programme of remedial works within the notice period, otherwise they run the risk of the landlord terminating the lease and seeking damages for all of the breaches. The extent of remedial works to be undertaken in such circumstances will vary in each case and will be a matter of balancing the degree and severity of remedial works and cost with the risk of forfeiture, damages and possibly business disruption if the works are ignored.

Section 147 relief against claims for internal decorations

If the landlord's section 146 claim includes claims for internal decorative works, the tenant may, subject to the statutory conditions being met, apply to a court to seek relief against claims of internal decorative works under section 147 of the Law of Property Act 1925.

The merit in seeking court relief under section 147 will be dependent on the extent and value of the internal decorative remedial works requested by the landlord. A pragmatic and sensible approach should be taken and costs of the action versus undertaking some or all of the decorative works should be taken into consideration.

Challenging alleged "damages" claims

Where the landlord serves a section 146 notice seeking forfeiture or to exercise a right of re-enter, they will normally have a right to seek end of term damages from the tenant if they successfully execute their notice actions.

On the occasions where the landlord fails to successfully execute their notice actions (such as when the tenant successfully claims relief from forfeiture under the Leasehold Property (Repairs) Act 1938, then it may become difficult for the landlord to then seek damages.

A key argument against damages during the terms is related to the basis on which the tenant pays rent. During the term, most tenants will be obliged to pay rent at a level initially set by the lease and thereafter reviewed and revised at set intervals. The regular "rent reviews" will commonly be undertaken based on comparable open market rents at the time of each review, and will disregard the impact on rents of any disrepair, ie the revised rent will assume a property is in full repair and good condition.

If the landlord is seeking the remedy of damages without the lease having determined, then it may be difficult for the landlord to demonstrate that they have suffered any damage where rents continued to get paid on a basis that disregard the negative effects of property disrepair. Under such circumstances, the tenant can usually demonstrate that the landlord has, in practice, suffered no loss and so there are no damages to pursue. However, such defence arguments will not extend beyond the end of the lease.

Landlord's "self-help" action

Understanding the landlord's tactical objective

If the landlord serves a "notice of repair" on a tenant under the "self-help" provisions of the lease, then the landlord has commenced an action that may ultimately result in dilapidations remedial works being carried out by the landlord, with the tenant then meeting the costs incurred as a commercial debt (see also Chapter 6).

The value to a landlord in implementing such works is that under the current law, the sums recoverable as a debt are not subject to the common law or statutory restrictions of loss that apply in compensatory damages claims. The other benefit to a landlord in undertaking such action is that, because the notice is not under section 146 of the Law of Property Act 1925, the tenant is not afforded any right of relief from the action under section 147 of the Law of Property Act 1925 or the Leasehold Property (Repairs) Act 1938.

A well advised landlord will be aware of the benefits of self-help action. It is not uncommon to find that self-help actions are commenced at strategic points in the lease as a means for the landlord to seek to exert negotiating pressure on the tenants on entirely separate sets of negotiations, such as rent reviews or for negotiating a surrender of the lease.

It should not be forgotten that self-help action is not without its risks to a landlord as it will require the landlord to "run-the-gauntlet" and to forward-fund often costly works. The landlord may then have to rely on further debt recovery action that could see costly and long running court challenges and disputes which carry a risk of less than 100% recovery if the landlord has acted beyond the lease. It remains relatively rare, therefore, that a landlord will actually pursue self-help action to the point that they actually undertake the works; although it remains a possibility in every case.

Notice conditions review

If a notice to repair is received, then the tenant's surveyor should carefully examine the self-help clause within the lease and ascertain the conditions on time available for the tenant to comply with the notice. Often, such conditions will state that the works are to either be commenced within a certain timeframe such as two months and thereafter prosecuted diligently; or may require the works to be completed within a set period.

Tenant response tactics — bluff calling

It is not uncommon for tenants receiving notices to repair to simply disregard the notice and in effect call the landlord's presumed tactical service bluff. But this is a risky strategy to adopt and is one that could well backfire if the landlord is well advised and genuinely intends to pursue the self-help action diligently.

More often than not, the tenant will seek to test the landlord's remedial works intentions and will endeavour to ignore the notice for as long as possible (until there are signs of genuine intention such as further surveyor return inspections, etc). There will however come a point where the cost risks start escalating where, if the tenant is not careful, they will find their only means of avoiding possibly significant costs will be to seek an injunction or to commit to undertaking at least some of the works.

Tenant response tactics — injunctions

The tenant should take into consideration that their rights of "quiet enjoyment" of the demise contained within a lease will ordinarily be conditional on the tenant not being in breach of the lease.

If the tenant is in breach of their lease in the manner indicated by the landlord's schedule of dilapidations, then they would not normally be able to prevent the landlord from temporarily entering the property to execute the self-help remedial works in keeping with any express lease provisions. This may prove both costly and very disruptive to occupation or business.

However, in some circumstances, it may be possible for the tenant to consider seeking an injunction on the landlord, preventing the landlord's access and entry to the demise property and the undertaking of works. Injunctions may be possible where the nature of the remedial works does not require urgent attention in order to safeguard the integrity of the structure and fabric of the premises; and where the landlord's actions would result in disproportionate or unreasonable disruption to a tenant's enjoyment of the property. Injunctions may also be granted where the landlord's intended works exceed the scope of works expressly permitted.

The tenant's surveyor should advise the tenant of the commercial debt cost risks if they disregard the notice to repair but it remains the tenant's prerogative to do so.

Tenant response tactics — limited works undertakings

Where it appears that the landlord is genuine in their intention to undertake the self-help works, the tenant should consider undertaking at least the most vital or urgent works within the self-help clause timescale. In doing so, the tenant will remain in control of tender and project costs that might be greatly favourable to the landlord's, no doubt, "Rolls Royce" standard works costs.

If the most vital works are addressed, then often the landlord will take a more considered and pragmatic approach to any remaining wants of remedial works. If the landlord remains dissatisfied, then the tenant could seek to negotiate a longer timeframe for compliance with the less urgent works.

The undertaking of partial or full compliance works programmes by the tenant is likely to prove effective, as the landlord, in most cases, would prefer that the tenant incurs works expenditure in the first instance.

Further enquiries

Landlord's surveyor conflict checks

The tenant's surveyor should undertake their own conflict of interest checks at the commencement of their appointment (see above). In addition, the tenant's surveyor should also carefully consider whether or not the landlord's surveyor may have any possible conflict of interest that could influence their handling of the claim.

If they have not already done so, the landlord's surveyor should be asked to clarify the extent of any potential conflict of interest they may have that could affect their handling of the case, such as other business relationships with the landlord or a performance-related or scale fees basis of appointment, etc.

When making any requests for clarification of possible conflicts of interest; and so they are not perceived to be hypocritical, the tenant's surveyor should openly disclose any interest that they have that could be considered to represent a conflict.

Where conflicts of interest are identified, the party with the conflict should be asked to consider, if it remains appropriate, their continued involvement with the claim and if they feel they can fairly attend to the claim without detriment to the non-conflicted party. Many current observers suggest that it would be risky for the conflicted individual or practice in question to remain involved and no doubt the courts will take any conflict of interest into consideration when considering conduct and costs.

Checking the schedule of dilapidations

The tenant's surveyor should arrange to inspect the property at the earliest opportunity after receiving the landlord's schedule of dilapidations.

The tenant's surveyor should seek to review and, most importantly, critically appraise the landlord's schedule of dilapidations contents, having regard to their own opinion of the tenant's obligations under the lease or any tenancy. The tenant's surveyor's inspection should consider the same issues described in Chapter 7.

It should not be forgotten that the tenant's surveyor's duty is not (as commonly assumed) to identify negotiating positions to tactically or knowingly reduce a claim beyond the fair and legitimate level. The tenant's surveyor owes their client a duty of care and skill in providing their services and their objective should be to advise their client from the outset of what their impartial opinion is of the true and fair claim. It should also be remembered that if the claim proceeds to court and the tenant's surveyor is to appear as an "expert", then their duty will be to inform the court of their true opinion of the claim. Any contradiction with earlier "negotiating position" representations may fatally undermine a defence case and call the surveyor's credibility into question.

Reviewing the landlord's schedule costings

Where a landlord's claim includes a claim for a monetary sum (such as when seeking damages attributable and arising from breaches of the lease), the tenant's surveyor should seek to clarify on what basis the claim has been valued.

With regards to the quantification ("costing") of the contents of the schedule of dilapidations, the surveyor should seek to ascertain if the costings have been based on "surveyor estimates" (and if so what was the source); or if the costings have been gained via a quantity surveyor; or from a contractor quoting for the works within a fairly run competitive tendering process.

If the schedule of dilapidations costings are based on a contractor quote gained under competitive tender conditions then, so long as the tendering process was properly and fairly conducted, these costs would be considered to be *prima facie* evidence. Costings based on any other estimation process will be more open to challenge and dispute.

Where costings are alleged by the landlord's surveyor to be based on costs from "similar" past projects, any such claim is a representation of fact and must be true. In these circumstances, a landlord's surveyors should be asked to openly disclose the similar past evidence and failure to do so may undermine the landlord's surveyor's credibility and severely damage the landlord's claim prospects.

Where costings are alleged to be based on published cost data books, such as the BCIS *Dilapidations Price Book* or the *BMI Building Maintenance Price Book*, then the landlord's costing appraisals should be carefully reviewed, as it is common to find that the price data has not been correctly applied or adjusted properly. Even if the price book data has been correctly applied and adjusted, it should never be treated as definitive as the price data is for an average price where the books do not publish the statistical population studied for each item and do not detail the population range or statistical standard deviation of the data. Typically, the difference between tender figures and the common price data books can vary by at least +/- 20%, so there will always be room for claims based on these "price" data books to be fairly contested.

Where a landlord's costings are based on mere speculative surveyor "gut-feeling" estimates, then there is a risk that the claim has been "recklessly" prepared. In these circumstances, the tenant's surveyor should form their own opinion and should look at other more suitable reference sources such as past projects or price data books.

When considering the costs claimed, regard should also be had to the landlord's stated intentions. If the landlord has stated that it is their intention to undertake the works but has provided a claim based on costs other than those gained from a tendering contractor, then the tenant's surveyor should seek to clarify this apparent discrepancy and request that tender figures are made available. It would then be difficult for the landlord to refuse the request without calling their statement of intentions into question. In these circumstances, the tenant may even wish to defer attending to or resolving the claim until such times as the landlord provides tender costs evidence that prove the genuineness of their intentions.

Checking the VAT "elected" status

The ability for a claim to include a claim for Value Added Tax (VAT) related loss or damages will be dependent on whether or not the property has been "elected" for VAT as this will determine whether or not the landlord can recover any VAT incurred in relation to the dilapidations claim (see Chapter 10).

A good starting point for reviewing the elected status of the property is to check whether or not VAT was charged on the last quarter's rent. If VAT was charged, then it is probable that VAT will be recoverable by the landlord and should therefore not feature within the landlord's claim.

Where there is any uncertainty with regards to the VAT elected status for the subject property, the landlord, or better still, their accountant, should be asked to openly clarify their elected status and if they are elected and can recover VAT, at what percentage rate.

From a tenant's prospective when receiving a claim they should be aware that they cannot force a landlord to opt to charge VAT on the landlord's building just to save the tenant paying it. The option to waive the exemption for VAT on any building is entirely a landlord's prerogative. Consequentially, if the landlord is not elected or opted for VAT on a building but the tenant is VAT registered, it may be more beneficial from a tax perspective if the tenant undertakes the works and is able to recover the VAT element, rather than pay it over to the landlord as compensation payment that they can't offset in their tax accounts.

Checking the claim value

Where the landlord's claim includes a claim for damages and once all other aspects of the landlord's claim have been checked, tested and examined, then the landlord's appraisal of the claim should also be critically appraised.

The landlord's heads of claim and statement of claim should be examined and tested on elements such as:

- claim items that may be rendered "valueless"
- claims for loss of rent, rates and insurance premiums
- claims for professional fees
- claims for a landlord's "management time" costs
- claims for waste
- claims for loss of amenity
- claims for VAT.

The tenant's surveyor should also seek to consider the landlord's intentions (and evidence of genuine intentions) so that proper regard can be given to any possible restriction or "cap" that could be applied to the claim under statute or common law.

Further information on the consideration due on the above issues can be found in Chapter 10.

Once the claim has been appraised, the tenant's surveyor should be in a position to advise their client of their true opinion of the actual loss (if any) suffered by the landlord and will be in a position to prepare a suitable response to the claim.

Obtaining declarations of truthfulness

Any claim made by a landlord should, in essence, be truthful, accurate and fair; and it would be unreasonable for the tenant and their advisors to be expected to respond to any claim made on a less than truthful basis.

PLA "endorsement" or statement of truth

The current Property Litigation Association's (PLA) *Dilapidations Protocol* suggests that the landlord's claim is accompanied by a surveyor's "endorsement" that the claim represents the true loss suffered. However, the surveying profession has taken issue with the PLA's proposals as they erroneously assume that the surveyor always acts alone in preparing the claim. The surveying profession understandably feel that this type of absolute declaration remains for the claimant (the landlord) alone to be making. Surveyors within the specialist Royal Institution of Chartered Surveyors (RICS) Dilapidations Forum have been actively promoting an alternative position where they feel that the landlord's claims should be accompanied by a surveyor's/advisor's statement of truth, drafted in the same manner as that expected of experts under the CPR Part 35.

Regardless of which type of statement eventually prevails and becomes accepted best practice (the authors prefer the CPR Part 35 style statement of truth), all landlord claims should be accompanied by a suitable statement concerning truthfulness.

Seeking a statement

Where a claim presented by a landlord or their surveyor lacks a statement of truth, then the tenant's surveyor should invite the landlord and their surveyor to make a suitable statement of declaration of truthfulness (such as a PLA endorsement of loss or a CPR Part 35 statement of truth).

Any declaration of truthfulness should be openly issued and signed by both the landlord, the landlord appointed surveyor and other advisors such as valuers who may have contributed to the preparation of the overall statement of claim.

Implications of no statement

If a landlord or their appointed agents refuse to make a suitable declaration, then the whole of the landlord's claim and conduct will understandably come under suspicion. In such circumstances, it may become impossible to make progress in resolving the claim and dilapidations dispute.

Where progress becomes frustrated due to a refusal to provide a statement of truth, the landlord will be left with no option to litigate if they wish to obtain a settlement, at which point they will have to make a statement of truth to the court. Consequently, there is no good reason for a landlord or their appointed advisors from refusing to provide a statement of truth right from the start and a refusal to do so may well have cost implications when the claim is eventually settled.

Reciprocal tenant statements

There is an old saying, "What's sauce for the goose is sauce for the gander", meaning if something is good enough for me, then it is good enough for you. Consequently, the tenant, their surveyor and other advisors such as valuers, etc should also be prepared to make reciprocal declarations of truthfulness when making their response to the claim.

Preparing the schedule response

Since the 1920s, it has become commonplace and best practice to submit any response to a landlord's schedule of dilapidations claim in a "Scott" schedule format (see Chapter 9 for an example of a Scott schedule).

Unfortunately, the purpose and benefit of utilising the Scott schedule format is not often understood and so the schedule responses provided by a tenant's surveyor sometimes achieves little in terms of rapidly crystallising the differences of surveyors' opinion. As such, it can hamper the prospects of achieving a timely resolution of the dispute and claim.

The benefit of a completed Scott schedule

Once the Scott schedule is properly and fully completed then both parties to the dilapidations dispute and their surveyors will have effectively set out the main arguments of claim and defence. Points of agreement can be set aside and efforts can then be focused on the disputed parts of the claim.

Where disputes remain, parties can prioritise and focus resolution negotiation efforts on the items of significant value; or alternatively can consider the benefit of pragmatic resolution of minor differences or even acceptance of disputed points if they are of relatively little importance or value. Any further agreements reached should be noted within the Scott schedule.

If the parties find they can't reach a complete settlement and that the points of difference are too great to resolve without court intervention, then they are able to commence formal litigation proceedings promptly and the completed Scott schedule will then become a prime document for the court to consider when determining the case.

The benefit to be gained by having a properly and honestly completed Scott schedule cannot be underestimated and it is a vital step in achieving a timely, cost proportionate and effective resolution of the dispute.

The Scott schedule completion process

The landlord's schedule claim

When the landlord's schedule is served it should contain a list of items in which the breach is described, a remedy proposed and, in damages claims, should also contain a single cost figure for the value of undertaking the proposed remedial works. This single figure is entered in the "the landlord's item, landlord's costing" column of the Scott schedule.

The tenant's valuation of the landlord's claim

The first step the tenant's surveyor should undertake when responding to the landlord's claim, is to complete the "landlord's item, tenant's costing" column of the Scott schedule. This column does not concern the tenant's surveyor's alternative opinion of any alleged breach, remedial works or costs associated with those alternative opinions. This column is solely for stating alternative costs estimates for the landlord's breaches/works claim expressly as presented.

The intention of this column is that it will allow the court to consider which of the two alternative opinions of costs and damages they will award for a schedule item if they uphold the landlord's allegations of breaches and appropriate remedial works when determining a claim. The completion of this column therefore presents a vital opportunity for the tenant to defend the claim by proposing possibly more reasonable costs where the likelihood is that the allegations of breach are true or will be upheld.

The tenant's defence of the claim

The next step is for the tenant to submit their opinion of each schedule item of the landlord's claimed breach and remedial works within the "tenant's comments" column. This is where the tenant's surveyor may state their alternative opinion as to the extent and degree of any breach (if any) and what the more appropriate remedial works are (if any).

The tenant's comments will form the basis of any future tenant's defence argument in court. Therefore, it is very important that the comments are the product of careful and impartial consideration. The tenant's surveyor should seek to remain reasonable and as fair and honest as possible when making their comments.

The tenant's defence value

Once the tenant has stated their defence comments and opinions, the tenant should then value their alternative opinions and comments and place their alternative value within the "tenant's item, tenant's costing" column of the Scott schedule.

Once the tenant's responses and values have been entered, the Scott schedule should be returned to the landlord for final completion and then re-issue.

The landlord's comment on the tenant's defence

The final step in completing the Scott schedule will take place once the tenant has issued their response/defence version of the Scott schedule to the landlord (see below).

Upon receipt of the tenant's completed Scott schedule response, the landlord's surveyor will complete the "landlord's comments" column of the schedule in which they make any further response they require to the tenant's defence comments contained within the tenant's comments column.

This is not intended to be an opportunity for the landlord's surveyor to "rant" or substantially revise the original claim; but merely to clarify, where necessary, if any part of the tenant's response comments are accepted; or alternatively, to assist the court by providing further materially relevant information in support of their original claim items.

The landlord's alternative defence value

Finally, the landlord's surveyor then sets aside their opinion of the claim, considers the tenant's comments and alternative opinion. They then value the tenant's defence position and place the values in the "tenant's item, landlord's costing" column of the Scott schedule, thereby completing the schedule.

Scott schedule costings check

Due to how the responses and values are submitted by the two surveyors completing the schedule, the only time where costs will be the same in all four columns for any given item is when there is complete agreement between the surveyors for the breach, remedy and cost.

The only time there will be identical costs in both the landlord's and tenant's item columns will again be whether there is agreement between the surveyors on a particular side of the argument.

For example, the tenant's surveyor may concede that if the landlord's surveyor's opinion of breach and remedy is correct then the landlord's cost is also correct. However, this does not mean that the tenant has accepted that the landlord's surveyor's claim is correct, and the tenant's surveyor is still at liberty to state their defence and maintain that it is the more appropriate schedule entry and cost if they believe it to be so.

Pre-litigation tenant responses

CPR Protocol

In theory, the structure of the Scott schedule document accords with the PLA Dilapidations Protocol and the binding CPR default protocol. The basic format of the typical schedule is considered to satisfy both the PLA version of the binding CPR default protocol and its derivative, the PLA *Dilapidations Protocol* of the tenant response document (with a series of columns to the right hand side of the landlord schedule), which evolved from the 1920s on the basis of dilapidations claims that went before trial judges, and on the basis of legal opinion on the clearest way to present the views of the respective parties to the dispute.

The PLA Protocol sets out a sensible timetable to follow in lease end dilapidations claims and requires that the tenant responds to the landlord's claim where possible within 56 days. This is intended to act as a brake on litigation where landlords might otherwise be overly keen to litigate for damages. Thankfully, dilapidations do not actually suffer this type of "hair trigger" approach to litigation and the mechanisms set out in the PLA Protocol are not addressing a particular need. Nevertheless, the approach set out in the PLA Protocol accords with the current and previous RICS guidance notes on dilapidations and is a sensible guideline.

Documenting the response

In response to the landlord's schedule, it is vital that the tenant's response is placed on the record at the appropriate time. Ideally, correspondence should not be "without prejudice" unless an offer of settlement is put forward. The reason being that the tenant will need to rely on any arguments put forward in relation to the liability and the information derived as a result thereof.

The label of "without prejudice" can have an unnecessarily confusing impact on the evidence that is disclosed in the initial stages of the dispute. Some surveyors incorrectly put the "without prejudice" wording on every letter, e-mail or fax sent to the other side. This is due to the lack of understanding by surveyors as to the use of that wording and can be used against them when they try to adduce evidence at a later date, which mistakenly has that wording on it.

A surveyor should look at the purpose of each piece of written evidence and ask themselves: what is the dominant purpose of this communication? Is it to raise a question that would assist in understanding the landlord's intentions (where "without prejudice" would be misused) or is it, say, to make an offer to settle on part, or all, of the claim (tenant's advisor); or to propose a sum that would be acceptable for all, or part of, the claim (landlord's advisor)?

A surveyor can also assess their opposite number's expertise in dealing with such disputes by the repeated incorrect use of this wording alone; and it is not unknown for advantage to be taken of that lack of knowledge when dealing with the negotiations of the case.

A clear and thorough approach by the tenant's surveyor will address any ambiguous matters as soon as possible and will show to the landlord and their surveyor that the claim is being scrutinised in the appropriate level of detail.

Client consent for settlements

It is vital that any counter valuation should be issued by the tenant's surveyor with the client's consent. This can be obtained in two ways:

1. At the start of the instruction the tenant surveyor should obtain written confirmation of instructions from their client, setting out the terms of the appointment. One important element of the terms of the appointment should be an unequivocal statement setting out that the tenant surveyor has an overriding duty to the court and that, as part of this duty, they will aim to come to the correct settlement figure in respect of the liability; and that this will entail open exchanges of correspondence between the surveyors defining and crystallising elements of the claim to the point where a total estimation of the probable liability will be set out in correspondence. This merely confirms in writing the legal duty and role of the surveyor, but crucially removes any cause for complaint that a client might have if they felt they were not involved in the exact quantification of the claim.
2. Alternatively, the tenant surveyor can issue the draft counter valuation to their client for approval to issue to the landlord. It must be explained clearly to the tenant client that the counter schedule is being prepared on the basis of independent expertise and that the counter valuation cannot be merely a figure that the tenant client wants to insert. In summary, the tenant surveyor must not allow themselves to become the mouthpiece for their client. Impartiality and reasonableness is key at all stages.

The counter valuation should be considered as distinct from the counter offer which is generally a without prejudice offer of settlement. The tenant surveyor must obtain their client's express permission to issue such an offer on the basis that such an offer, if accepted, will form a binding agreement settling the claim. Any surveyor issuing such an offer without express consent from the client will risk a negligence claim if the client is not happy with the agreement.

13 Negotiation and Settlement

Stage 1: Negotiations

Duties of negotiators

The courts have given consideration to the duty of parties performing an advocacy "negotiation" role where negotiations concern matters requiring utmost good faith between the parties. Case judgments such as that of Millett J in *Logicrose Ltd* v *Southend United Football Club Ltd (No 2)* [1988] EGCS 114 have confirmed that: "Parties to negotiations do not owe each other a duty to act reasonably, but only to act honestly."

It could be said that the nature of dilapidations claims are claims where representations made during the course of a claim should be capable of being taken in utmost good faith. Any settlement negotiated and agreed will, in itself, constitute a legally enforceable contract between the parties. Consequently, landlord and/or tenant parties together with all their advisors engaged in negotiations (including surveyors) should conduct themselves in a totally honest manner at all times.

Any party or advisor who fails to act honestly will not only call their own credibility and case into question but could even find that they have committed an offence (see Chapter 14).

Surveyor negotiations

Surveyors are very good at settling disputes. The best evidence of this is to consider how few dilapidations disputes actually result in proceedings, and in turn how many of these end in a trial. Only a very small percentage of these claims actually litigate, however, some surveyors say this is down to the general lack of knowledge and reluctance to become involved in legal proceedings.

This is not a new post Civil Procedure Rules (CPR) view of litigation probabilities. Well before the introduction of the CPR, surveyors formalised an approach of quasi alternative dispute resolution (ADR) whereby two "advisor" professionals representing either party in the dilapidations damages claim commenced a process of disclosure of their respective evidence and arguments. The process is of course more commonly known as the schedule of dilapidations served by the landlord and the tenant response document, which has become commonly known together as the "Scott" schedule (see Chapter 9).

Once the schedule of dilapidations has been served, the surveyors should aim to meet on site to jointly assess the breaches alleged to allow the tenant to record them in the Scott schedule format.

A diligent approach at the start of the claim, combined with a pragmatic approach to the negotiations, will facilitate the fair settlement of the dispute. This is ultimately the goal of most parties to the dispute and by adopting this approach the surveyor is more likely to have a client that is pleased with the services provided and, crucially, the surveyor will also be in compliance with the CPR.

Pre-litigation conduct and the CPR

Contrary to popular belief, the CPR cannot be ignored until litigation is imminent. All negotiations must be conducted within the strict parameters of the CPR. The principal rationale of the CPR was to settle disputes without recourse to litigation. This has resulted in a set of procedural rules that focus the parties of methods of settlement that include:

- Part 36 offers
- exchange of information at the start of a dispute
- settlement meetings
- ADR
- responses by the defendant within reasonable time limits.

The environment created by the CPR creates a system where claims tend to settle.

Until the Property Litigation Association's (PLA) *Dilapidations Protocol* (originally produced back in 2000) is approved by the courts (if it is ever approved), the present protocol to be followed during the pre-litigation stages of a dilapidations claim is the protocol within paragraph 4 of the CPR Practice Direction on Protocols (the Default Protocol).

The CPR Default Protocol is the significant means by which the CPR's overriding objective is achieved. It sets out a broadly applicable programme for the resolution of disputes relating to allegations of disrepair of commercial premises at the lease end and other consequential disputes that occur between landlord and tenant.

The CPR Default Protocol is a binding legal obligation and is suitably non-prescriptive in terms of timescales for service of a landlord claim and makes useful reference to the appointment of experts at paragraph 4.9, even acknowledging that joint experts are not going to be useful in all cases, which is probably often the case in a dilapidations claim.

The latest version of the Royal Institution of Chartered Surveyors (RICS) *Dilapidations Guidance Note* (5th ed, June 2008) has finally caught up with the CPR and now refers surveyors to the protocol within the body of the guidance and includes an extract from the CPR Default Protocol in appendix G.

If the CPR Default Protocol is read in conjunction with the RICS Guidance Note, surveyors and lawyers have all that they need to address most pre-litigation issues relating to dilapidations dispute resolution in compliance with the CPR. Their actions will be legally appropriate and will adhere to the basic principle of the CPR protocols, ie encourage the free flow of information and prevent litigation. There are also costs and court "sanction" risks if the CPR Default Protocol is not followed.

Meetings between negotiators

It is advisable that meetings should be held between the parties and/or their advisors at an early stage in the dispute. Further meetings should be considered as often and as necessary thereafter in order to achieve proportionate but beneficial progress towards resolving the dispute.

If beneficial progress is to be made at a meeting, those attending the meeting should clarify the extent of any documents, evidence or statements that they require (within reason) in advance of the meeting so they have an opportunity to give due consideration to the relevant issues prior to the meeting. It is then advisable that both sides seek to provide and disclose the requested information to their counterparts in the spirit of co-operation so the meeting can proceed.

The meeting attendees should agree in writing in advance that the meeting will be conducted and treated as a "meeting of experts". For example, the meeting should be held on an agreed and acknowledged without prejudice basis to encourage frank dialogue and progress where the detail of meeting discussions remain without prejudice and inadmissible. Importantly though, the general extent and reasons for any agreements reached during a meeting of experts is "on the record" and is admissible. Similarly, the extent of any key points of disagreement (but not the fine detail reasons and discussion) will also be on the record.

If progress is to be made and if there are some difficult issues of differences of opinion between the parties then it is beneficial to agree a meeting agenda in advance on the meeting. The agenda can help maintain dialogue and progress and where discussions on any particular point become stalled due to irreconcilable differences, then they can be set aside and the next agenda item addressed (often after a short break to clear heads and cool tempers).

"Ambushed" meetings

The parties/agents attending the meeting should agree in advance who will be present so that "ambushed" meetings can be avoided where, say, one party unexpectedly turns up with a full legal team in tow and without having provided the other side with the opportunity to arrange for comparable professional attendances.

Where "ambushed" meetings occur, it is reasonable for the original meeting attendees to either agree that the meeting is re-convened to a later date once both parties have resolved the issue of who should be in attendance; or the unexpected attendees could be invited to leave. If the parties cannot agree who should remain at the ambushed meeting then it may be more prudent to terminate the meeting and provide a cooling off period before further meetings are attempted.

Surveyors who find themselves suddenly ambushed should only remain at an ambushed meeting with all unexpected attendees present if they have first gained their client's or client's solicitor's approval to do so.

Documenting the meeting

Where a meeting is held on a "meeting of experts" basis (see above) the "on the record" agreements reached or general reasons for continued dispute should be recorded and agreed in writing by those present, without interference or undue direction from the client's legal advisors (if they were not present).

The meeting record will be a document that is admissible into court at a later stage. The record is intended to form a truthful and open record of the meeting so if one side or the other seeks to suddenly go back on their meeting agreements then their conduct and credibility is open to question and further examination if they cannot show good reason for any change of position.

The meeting record can be in simple "minute" form or, in more complex cases, can be set out in a tabular format in which columns are provided for recording in general terms the issue raised; whether agreement was reached; and outline reasons for any continued disagreement or conditional agreements. Once prepared and agreed in terms of content, the record should be signed by those present and openly circulated.

Without prejudice correspondence

Correspondence marked "without prejudice" has to comply with certain rules. The professional issuing such correspondence must be aware of the legal nuances pertaining to this legal phrase, as it is so often misused in lease end dilapidations claims.

A document marked "without prejudice" is privileged from disclosure to the court in any proceedings between the author and recipient and cannot be used as evidence by either party, unless the parties agree to waiver that privilege. Use of the phrase enables one party to a dispute to put forward settlement proposals at no risk of subsequent adverse disclosure by the recipient of the offer.

The without prejudice communication must be made in the course of genuine negotiations to settle actual or contemplated litigation and not, as stated above, on every single communication sent on the dispute to the other side.

The rule stems from two sources:

1. The public policy requirement, going back hundreds of years that permits parties to a dispute to be able to seek to settle them,

without "prejudicing" their position, to facilitate settlement and compromise.

2. An implied agreement between parties.

Merely heading a document "without prejudice" does not automatically afford protection to the writer — a person who makes an admission of fact on a without prejudice basis in the belief that it cannot be disclosed, fundamentally misses the point and risks exacerbating an already heated situation. Without prejudice communication can relate to offers of settlement only. The "without prejudice" label cannot be used as a "cloak for perjury".

By way of clarification, Judge Walker LJ, in the Court of Appeal case of *Unilever plc* v *Procter & Gamble Co* [2000] 1 WLR 2436 set out the circumstances when evidence obtained in the course of without prejudice negotiations could be disclosed:

- When the without prejudice communications have resulted in a concluded compromise agreement.
- When it is argued that an agreement reached due to without prejudice negotiations was due to fraud or misrepresentation.
- If one party makes a statement upon which the other relies and acts upon, the other party may raise an objection under the principle of "estoppel".
- One party may be allowed to give evidence of what another said during without prejudice negotiations if the exclusion of that evidence would act as "a cloak to perjury, blackmail, or other unambiguous impropriety".
- Offers to settle proceedings "without prejudice except as to costs".

The case of *Dora* v *Simper* [2000] 2 BCLC 561 examines the definition "unambiguous impropriety", concluding that such a statement was admissible if the claimant stated that they intended to pursue a bogus claim or use fraudulent means. The case also stressed that admissions were disclosable, when they were not attempting to resolve the dispute, ie not part of genuine settlement negotiations.

In summary, the without prejudice rule generally results in communications between parties not being disclosable if they are made during genuine attempts at settlement. The area of without prejudice communications highlights the risks that are created when building surveyors advise clients on legal issues.

Stage 2: Settlement strategies

Avoiding wasted costs

One of the few certainties in commercial dilapidations disputes is that lawyers and surveyors will charge a fee for the work they perform for their client on the basis of their letter of appointment. This is an unpleasant reality for many clients and is often at the forefront of their minds during the instruction.

To avoid incurring wasted costs, both the landlord and tenant surveyor should address the claim proactively as follows:

- Set out the claim and response clearly in unambiguous terms.
- Arrange meetings between landlord and tenant surveyors as soon as possible and prepare for them fully.
- Carefully explain to the client the exact purpose of each item of key correspondence and the reason for any meeting. Ideally, the surveyor should have instructions to settle at a reasonable figure as soon as possible (in principle and subject to contract) and should be able to communicate any agreement in principle to the client efficiently so that the settlement of the claim is not unnecessarily delayed.

Early settlement proposals

Settlement payment offers

Based on the principle that a dilapidations claim is merely a claim for damages for breach of contract, the potential defendant in any litigation should always consider making an offer of settlement. This can involve the actual payment of money by one party to the other and should always be dealt with by a lawyer. In the context of dilapidations there are two types of payment options, both of which can incorporate a Part 36 offer to create extra focus on settlement:

Payments on account

Payment of a sum of money on account of any damages awarded can be a useful approach to adopt by the potential defendant to mitigate the defendant's liability for interest payments if the matter is litigated.

A typical example of a settlement payment offer would be where the defendant tenant surveyor had valued the landlord's claim at between £200,000 and £400,000, subject to disclosure of further information that could result in a valuation at any point on this band. The tenant can make an offer to pay a sum of money in respect of damages on account at, say, £100,000. The advantages of such an approach are as follows:

- It can pacify an aggressive and confrontational claimant landlord.
- It can offer protection in terms of costs.
- It can sometimes remove the threat of litigation altogether as the balance of the potential liability falls below the claimant's own threshold for litigation.

If there are parts of a dispute that are agreed, a payment can be made to settle just these elements, thereby avoiding the sums being considered as a payment on account for all of the potential liability.

Payment in full and final settlement

Alternatively, the offer can be issued unilaterally with a cheque issued without notice on the basis that it is offered in "full and final settlement". Psychologically, a person is more likely to accept an offer to settle if the payment is put in front of them and they are given the chance to gain physical possession of the money. This is often a far more tempting approach than merely issuing a written offer and waiting for a response.

Such an offer should be carefully worded and ideally be issued by a lawyer. The offer will need to be marked "without prejudice save as to costs" and will be capable of disclosure to the court if the matter litigates and the issue of costs or interest is debated.

This approach opens up the age old legal debate as to whether a cheque cashed by the recipient that is volunteered by the paying party in "full and final settlement of all liabilities", but in a sum less that the level claimed, constitutes settlement of the liability when the cheque is cashed. Unfortunately, there is no definite judgment on this issue but the following factors should be borne in mind by any claimant tenant seeking to adopt such an approach or any landlord receiving such a payment:

- The recipient must respond quickly if they are going to reject the settlement of the dispute and also cash the cheque as part payment.

- The burden of proof is on the paying party (the defendant tenant) to prove that there was an agreement reached. This tends to be based on the time elapsed between the receipt of the cheque, the cashing of the cheque, and the rejection of the "full and final" basis of the offer.
- If the recipient of the payment wants to treat the dispute as continuing and the payment merely as "on account", then the recipient needs to state this as soon as possible. Case law suggests that this can only be a matter of days rather than weeks, but each case will turn on its own facts.

When acting for the tenant, this approach should be considered carefully and used only when based on legal advice, with a client that fully understands the risks of the tactic. When acting for the landlord, this approach should be guarded against and if such an offer is received the response must be quick and unequivocal, ideally via the lawyers.

Part 36 Civil Procedure Rules 1998

The Civil Procedure Rules (CPR) comprises a total of 75 "parts" dealing with all civil litigation issues from traffic enforcement to trial costs. Amongst many areas of civil dispute, Part 36 relates to commercial property dilapidations.

Part 36 deals with offers by either side to settle the issues at stake by way of a written offer, which is automatically "without prejudice save as to costs". Part 36 is a very effective tool for either side involved in a dilapidations claim to use at any stage of a dispute by creating a risk for the recipient in relation to costs.

The offer has serious costs and interest penalties if rejected, if the party rejecting fails to do better at trial. Either party can make an offer at any stage, but importantly the offer can be made before the claimant (often the landlord in dilapidation disputes) issues court proceedings, ie it is useful right from the start of the dispute, potentially from the point that the schedule is served.

The situation pre-CPR 1998

Pre-CPR only the defendant (ie tenant) could make a "payment into court" (a formal offer to settle which had certain adverse cost and interest consequences if wrongly rejected). This was only possible after proceedings were issued, ie often once the dispute was at a fairly progressed stage, and certainly when lawyers were involved as the principal advisor and/or negotiator. This meant that previously it was a lawyer's job, allowing the surveyor to be relatively safe in blissful ignorance. Surveyors representing defendant tenants were able to make "Calderbank Offers" but there was no certainty in respect of costs based on the level of offer reference to the outcome at trial.

Background to Part 36 CPR 1998

Part 36 of the CPR was perhaps the single most important change to the procedural rules governing the pre-litigation conduct of civil damages claims introduced by Lord Woolf in 1998. The concept of legally binding pre-litigation offers with potentially post-litigation costs consequences is only rivalled in terms of importance to dilapidations dispute resolution by the CPR protocols; a particularly suitable comparison because these two elements of the CPR are so closely intertwined. With costs often running into tens of thousands of pounds for a claim of just £100,000, the implications of Part 36 are potentially serious.

Under the CPR, the surveyor representing the landlord or tenant has the opportunity (and arguably the obligation) to protect their client's interests by making an early Part 36 offer well before proceedings are issued — if circumstances are appropriate. Since April 1999, the CPR has specifically promoted such offers before cases are litigated, at a time when non-lawyers are often the principal advisor. Provided the clear rules of Part 36 are followed in respect of the terms of the offer, the tactic should be used by the dilapidations surveyor.

Therefore, there is no reason why surveyors cannot issue such offers at any stage of a dispute or, at the very least, make a recommendation to their client's lawyers that such an offer should be made if they feel unable or unwilling to draft the offer letter themselves.

Basic principle of Part 36 CPR 1998

Part 36 offers can be issued by tenant or landlord. The consequences of Part 36 apply if one party to a dispute makes a Part 36 offer and:

- the recipient does not accept it
- the matter goes to trial

- the party who did not accept the offer is awarded less by the trial judge than they would have received had they accepted the offer in the first place.

In the case outlined above, the party who did not accept the offer can be penalised in legal and other costs either by not being awarded some or all of the costs they would normally have been entitled to or by being required to pay more than they would otherwise have been obliged to do.

The penalty normally applies only to costs incurred after the period during which the offer could have been accepted, ie 21 days after it is received.

The concept of a Part 36 offer gives the landlord or tenant a tactic for promoting settlement at a far greater pace than was the case prior to 1998. The concept is best explained by way of simple example.

Example

If, for example, a landlord is claiming damages in respect of a lease end breach of covenant in sum of, say, £400,000, the tenant may offer to settle for, say, £300,000. The landlord may not be sure about how strong their case is, because litigation has not commenced and the formal process of disclosure and expert meetings has not taken place. The landlord may therefore be tempted to settle.

By virtue of a Part 36 offer, the landlord can issue a legally binding without prejudice offer for, say, £350,000. If the tenant refuses and the matter subsequently litigates and the tenant is ordered to pay more than £350,000, the court can conclude that the tenant should have accepted the offer and make an order that the tenant pays the landlord's costs, plus the interest on the sum found to be due at the rate of 10% above base rate. Such a punitive rate of interest can hugely increase the burden of any judgment, particularly if the Part 36 offer was made early on in proceedings with a substantial costs liability incurred by the successful party conducting the ensuing negotiations and litigation.

Practical aspects of Part 36 CPR 1998

The potential consequence of this mechanism is that the landlord can offer to settle without damaging their chances of maximising their prospects of recovery in the main litigation. In cases where basic liability is not in dispute but only the quantum of the claim, the claimant can be confident that they will recover their costs however small the amount they are actually awarded.

However, if the defendant makes a Part 36 offer, pitched at the right level, it can make the claimant nervous that they may not recover their costs if the eventual award is less than the offer, thereby shifting the litigation risk onto the claimant and promoting settlement.

Consequences of Part 36 CPR 1998

Partly as a result of the introduction of Part 36 offers within the CPR 1998, there was a substantial fall in the number of cases litigated. The new rules empowered clients and non-lawyer professional advisors (eg surveyors) to settle disputes themselves without litigation, thereby saving legal costs.

Terms of the Part 36 offer

The letter making the offer must comply with the following requirements (set out within Part 36.2):

- It must be in writing.
- It must state on the face of the offer that it is intended to have the consequences of Part 36 and state that the offer is "pursuant to Part 36 CPR 1998".
- It must state whether the offer relates to the whole of the claim, or just part of it. It is possible to settle item by item and this needs to be made clear. For example, in a dilapidation dispute, the costs for replacement carpets or redecorations can be removed from any potential litigation by way of a Part 36 offer on just that item, if accepted.
- It must specify a period of not less than 21 days within which the defendant will be liable for the claimant's costs, in accordance with rule 36.10 if the offer is accepted.
- It must state whether it takes into account any counterclaim.

While the offer is issued as a normal letter (not a court form), the author must be very careful that all issues are clearly stated because the document could prove crucial several years down the line. The offer should be posted and faxed, with the fax receipt kept on file.

Recent changes

Part 36 of the CPR was significantly amended in April 2007 when the Civil Procedure Rule Committee (CPRC) decided to adjust the procedure to make it more practical, quicker and flexible. The principal changes were as follows:

1. Payments into court were abolished:
   (a) Both before and after the introduction of the CPR 1998 defendants were required to make payments into court in order to gain costs protection once proceedings were commenced.
   (b) From 1998 to 2007 the CPR required the offeree (very often the defendant tenant in dilapidation proceedings) to back up a pre-litigation offer of financial payment with a payment into court to continue the protection on costs from the date of the offer.
   (c) Post April 2007 such offers simply need to be in writing to comply with Part 36.

   This change of procedure was to avoid the court holding substantial sums of money, often paid in by insurers reluctant to lose access to capital tied up for many years.

2. If the Part 36 offer is accepted, the sum offered must be paid within 14 days of acceptance:
   (a) The defendant must offer to pay the sum within 14 days of acceptance. If a defendant's or claimant's offer is accepted but then not paid within 14 days, the claimant can enter judgment immediately to conclude the litigation in the sum offered.

3. Withdrawal of offers:
   (a) Previously, the court's permission was required to withdraw or reduce a Part 36 payment into court. Following the abolition of payments into court, CPR now states that the offers can be changed simply by serving written notice on the other side. If an offer is withdrawn at any stage, it ceases to have any Part 36 consequences. If an offer is not withdrawn, a counter-offer and or rejection of the offer will not prevent an offeree from subsequently accepting the offer at any time right up until trial.

Is there a risk of professional negligence?

Despite the straightforward nature of the Part 36 offer, the level of understanding and utilisation among surveyors involved in dilapidation disputes remains very low. It is a powerful tool with potentially costly consequences if the offer itself, or the response, is poorly judged. The CPR has been with us since 1999, with widespread discussion on the changes to civil procedure since 1997. Surveyors involved in dilapidation claims representing either landlord or tenant parties have the opportunity, and arguably the obligation, to make these offers.

Summary

The strength of Part 36 in relation to dilapidations is that it gives a landlord or tenant the power to create real pressure to resolve a dilapidations dispute. An adverse order on costs by the trial judge is the principal risk that is faced by any party to a dispute receiving a dilapidations claim. Of course, costs are always at the discretion of the court so a party can never be absolutely certain that using the Part 36 procedure will have the desired result. However, Part 36 offers are a useful tactical weapon in dispute resolution at most stages of a dispute.

While some dilapidations disputes will always litigate, Lord Woolf's aim of reducing costs and speeding up the litigation process has directly impacted the role of the non-lawyer via Part 36. Therefore, the work of a surveyor is now even more closely linked to legal principles and procedure, particularly in areas such as dilapidations disputes. Action (and non-action) at the early stages of a dilapidations claim can have lasting consequences. Much of a surveyor's work is close to the inception of a dispute, often before the lawyer is instructed to examine the claim in detail and often before the client even realises there is a dispute.

"Wait and see" tactics

Advice in relation to the resolution of dilapidations disputes often focuses solely on the proactive measures that one set of advisors can utilise. These measures include the preparation and service of the schedule, pre-action protocols, Part 36 offers, and section 18(1) of the Landlord and Tenant Act 1927 valuations, etc.

An often ignored tactic in dilapidations dispute resolution is the "wait and see" reactive approach, which in some cases is a useful adjunct to the more commonly used methods and in some cases a "battle plan" all by itself.

For the tenant

It is impossible to list every example of where the "wait and see" tactic can be utilised. Examples of where the approach can pay dividends for a tenant can relate to situations where the landlord is attempting to fast track a claim towards settlement with tactics akin to a "bully boy" approach, eg where the landlord serves the schedule prior to lease end, claiming that the work will be undertaken and that they want a damages payment agreed immediately. The tenant should consider adopting the "wait and see" approach to ascertain whether the claim is actually valid.

Very often a landlord's true intentions will define the extent of the tenant liability, and will only be known once the landlord has physical possession of the demised premises after lease end. This approach allows the tenant to answer the questions: "is the landlord actually going to put his money where his mouth is?" and "will the actual works include any element of supercession?"

Landlords following the CPR Default Protocol are required to state their litigation and ADR intentions and so often threaten legal proceedings at the start of a claim. There are numerous procedures that landlord claimants must comply with prior to proceedings actually being issued, with the CPR protocol being perhaps the most important. Until these matters are dealt with the landlord is in most cases unlikely to litigate. Landlords will be as conscious of the costs of litigation as tenants and any threats of this nature are worth reviewing carefully with the lawyers to avoid a rushed tenant response.

For the landlord

Another example is where a tenant gives assurances to the landlord that the demised premises will be left in covenant compliance at the lease end date. Rather than enter into binding agreements or serve schedules of dilapidation prematurely, the landlord should carefully consider waiting to assess the true state of disrepair at the end of commercial leases.

Equally, the landlord should not be pressured into accepting a financial deal prior to the end of the lease, until the true costs of all necessary and relevant works are correctly priced. This can often only be fully ascertained once the premises are vacant.

If the tenant claims that they are about to start the works and that a cash settlement has to be accepted by the landlord "now or never", the landlord should consider the following:

- Advise the tenant that landlord consent/approval may be required in respect of certain matters of reinstatement, repair and redecoration.
- That planning consent and/or building regulations approval may be required.
- Suggesting to the tenant that if works are not completed in strict accordance with the lease obligations, then there is a risk of double liability to pay both their contractor's costs and landlord damages in respect of any proven breaches of lease covenants once the works are completed.

Summary

Because dilapidations disputes can be slow to resolve, it is important not to always fast track the settlement negotiations, despite the environment of prescriptive timetables for the conduct of the claim.

Proactive use of the CPR Default Protocol

One of the many unusual features of commercial property dilapidations disputes is that surveyors are the principal professional advisors in what is really a legal damages claim.

This current process has developed over many hundreds of years and there is no reason that it should not continue, provided that the client's professional advisors are fully up to date in respect of the relevant legal procedures as well as the technical surveying issues. Knowledge of the CPR 1998 Default Pre-action Protocol is essential if a surveyor is to resolve claims appropriately.

Lawyers and surveyors can work together to use the PLA protocol to encourage landlords to litigate in the face of unreasonable tenant procrastination. Turning the protocol on its head and using it to drive slow claims forward, is an alternative to endless negotiations sometimes faced by surveyors representing landlord clients, even when the works have been remedied at the landlord's expense.

An anomaly of the PLA *Dilapidations Protocol* from its inception back in October 2000 was that dilapidations disputes have traditionally never suffered the "fast track to litigation" problems that other areas of civil dispute witnessed. One just has to compare the low volume of dilapidations litigation over the last 10 years with the far higher volume of other types of claims, eg housing disrepair, personal injury and medical negligence. While dilapidations arguably did not need a specific protocol, ie one that is designed to avoid litigation, now that it exists surveyors might as well use it to its full advantage.

By virtue of the existence of the PLA Protocol, there is now a mechanism to change the way that some dilapidations claims are managed and settled to achieve the right result more quickly. If working for the landlord, lawyers, surveyors and valuers can use the PLA Protocol to the client's advantage because the Protocol sets out a very clear timetable towards litigation with some relatively straightforward hurdles prior to the issue of proceedings.

Once the Protocol has been complied with, why not litigate? If the client has been wronged and no reasonable offer is forthcoming then what else can a reasonable party do? Liability is very often not in dispute. If the landlord has suffered a loss due to the tenant breach and no Part 36 offer has been forthcoming from the tenant, then the landlord should consider pushing ahead to commence court proceedings.

A substantial proportion of commercial property dilapidations disputes in existence could be "fast tracked" towards litigation by better strategic and tactical claims handling on behalf of the claimant landlord. This may lead to a quicker and fairer result for landlords, whose legal team would be re-paid elements of their fees on the recovery.

Because too few dilapidations claims are currently litigated, tenant surveyors are generally aware of this and behave accordingly with no incentive to offer a reasonable sum in settlement.

Using the dilapidations protocol as the basis for the landlord's timetable, the landlord should consider a strategy as set out in the flow chart on p202.

Documenting the settlement

Once a settlement has been reached, the parties can agree to document this by a simple exchange of open correspondence (ie not marked

Schematic showing how the proactive use of the PLA Dilapidations Protocol can expedite settlement of landlord claims

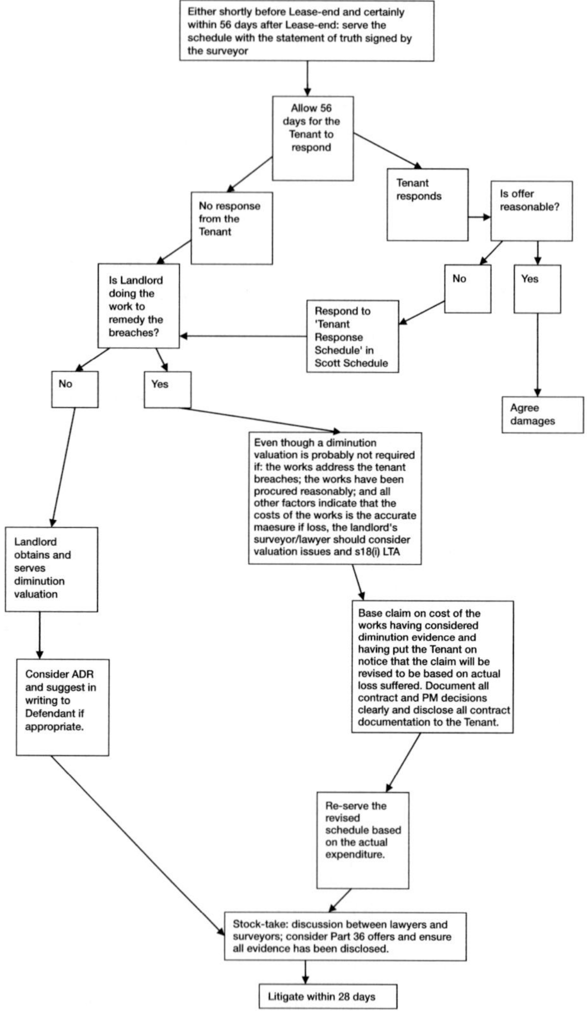

without prejudice) or may decide to have this documented by solicitors preparing a formal document.

There are pros and cons of each method. Clearly, the simple exchange of correspondence by the surveyors is likely to be the most cost effective, but the formal document may be appropriate if there are any particular issues that need careful consideration in relation to the documenting of the settlement or eg there are timing requirements for the payments of the monies that need the warning of sanctions for delay in payment.

Financial Services Authority checks

Whenever monies are to be paid to a party, it is advisable that full checks are made as to the identity of that party. It is not unknown for a party to use property dealings as a "front" for illegal activity involving money laundering.

On the basis that some dilapidations claims involve many millions of pounds, this may prove a magnet for some parties wishing to pass monies obtained illegally to another party without much attention from the authorities.

Solicitors and surveyors may become unwitting participants in a plan to pass monies over from a "tenant" to a "landlord" by way of a dilapidations claim of millions of pounds. Where solicitors and surveyors are acting, some individuals may consider this presents an opportunity to "clean" monies obtained from illegal activities.

Unless advisors know the true identity of those for whom they act, they can face not only embarrassment, but also criminal charges under various statutory provisions such as the Proceeds of Crime Act 2002, the Fraud Act 2006 and the Money Laundering Regulations 2007 to name a few.

Checks on a party's identity may be made via the Companies House's website (*www.companies-house.gov.uk*), via the Land Registry's website (*www.landregistry.gov.uk*), while Dun & Bradstreet, Experian or ICC can provide credit checks on parties.

Stage 3: Alternative dispute resolution

Alternative dispute resolution, often abbreviated to ADR, is the collective name given to the methods available in order to settle a dispute that does not involve the litigation process. They are generally considered to be:

- mediation
- arbitration
- independent expert determination
- early neutral evaluation.

Perhaps surprisingly, negotiation between the parties is classed as a form of ADR as it is a process that is designed to settle disputes that does not involve court procedures.

Some of the ADR procedures are binding on the parties, to the extent that there is a settlement reached that cannot normally be referred to a court by a party whom is not happy with the outcome; and some are not binding. Some ADR procedures may be used in conjunction with others (for example, mediation followed by arbitration) and some can only be used on their own, eg if parties agree to arbitration, that would create a binding decision so that other forms of ADR cannot be used.

In order to assess the pros and cons of each form of ADR, it is useful to look at each in turn.

Mediation

When parties to a dispute think about ADR, mediation is the process they think about more than the others. The reasons for this are that there are distinct advantages with mediation over the other forms and that is why it has enjoyed the prominence that it has.

One of the main advantages of this type of ADR is that the process is not binding on the parties so there is a freedom to go through this process knowing that if either party is not happy with the position reached after the process has been exhausted, then another ADR process may be proposed or litigation may commence. There is generally no disadvantage with mediation if it is conducted correctly with a suitable mediator.

In a nutshell, the parties agree (or are ordered by a court) to appoint a third party mediator who would be suitable to mediate the dispute. With service charge and dilapidations disputes, the issues can only be amount sums of money that should be paid from one party to the other and so the matter in dispute is clear. When mediation is used to settle other types of property disputes, such as the correct position of a boundary between two properties or where it is used for non-property disputes (for example family disputes), the matters are more complicated.

Each mediation will be conducted in different ways, depending on what the parties agree or what the mediator finds the best method. However, a typical mediation will run as follows:

1. The mediator will ask the parties to prepare a position statement which sets out each party's final position of the dispute following the negotiations that will have taken place previously. The mediator will meet the parties together at the start of the mediation with their legal and surveyor advisors if they have instructed them in order to inform them of what will happen on the day.
2. The next stage is for the mediator to meet the parties separately to assess their position with them against the other party's position. The mediator will discuss likely settlement positions with each party in their separate rooms with the intention that the parties will be brought closer together. The parties then meet together with the mediator to assess their respective positions which may lead to a settlement at that stage or may need a further round of the mediator speaking to the parties in their separate rooms again.

Studies show that most mediations lead to a settlement and the costs of the process are generally lower than the other forms of ADR and litigation. The only costs are the professional fees of the mediator, the hire of a suitable venue with the series of rooms that are required and the advisors attending the mediators if that is what the parties wish. Clearly, if the parties themselves meet without their advisors, the costs will be lower.

Another reason for the popularity of mediation is that the High Court and the county courts have the power to award costs against a party whom has unreasonably refused to consider ADR. Even when litigation has started, it has been known for the judge to inquire, when at the pre-trial review stage, whether there would be any benefit in the parties mediating their dispute prior to the court setting down a timetable for the exchange of witness statements, the disclosure of documents and the other matters that need agreeing before a trial date.

Arbitration

With this form of ADR, a third party is appointed by the parties in order to settle their dispute. However, whereas mediation is not binding on the parties and that third party is only engaged to facilitate a settlement, an arbitrator makes a final, binding decision, known as an award, based on the evidence given to them. The arbitrator can only rule on the dispute within the range of sums passed to them from the parties and not outside of that range. For example, with a dilapidations dispute, the tenant or former tenant may wish to offer no higher than say, £80,000, whereas the landlord wishes to obtain £200,000. The arbitrator cannot award a sum lower than £80,000 or higher than £200,000.

As the arbitrator's award is final and the sum awarded has to be within the range of the difference between the parties, there is little difference between this type of ADR and litigation. There are very limited grounds for challenging an arbitrator's award: where there has been a serious irregularity, a challenge to the arbitrator's jurisdiction or a point of law has been misinterpreted, a party may refer the matter to the High Court for a determination. There have been very few property cases that have been challenged at court and even fewer which have been successful.

The only advantage over litigation is that usually the arbitrator appointed is a specialist in that field of the dispute and not a county court or High Court judge who would not have practiced in that field. That specialism leads to a saving in costs as there should not be the need for additional expert witnesses to be employed to assist the arbitrator in determining the issues, unlike at court, where experts are often required to give their opinion on the facts in dispute.

However, it may be the case that costs of arbitration may be higher than litigation, due to the arbitrator normally charging on an hourly basis at, say, £300 to £400 per hour; whereas court fees may be lower, as these are fixed sums merely based on the sums in dispute. Therefore, where a case is likely to take a number of weeks, it may be less costly for it to be settled at court rather than by an arbitrator charging on an hourly basis.

However, where the dispute is required by the parties to be settled as soon as possible, an arbitrator would probably be able to deal with the case quicker than a court where it usually takes over six months for a trial to be heard at the High Court. An arbitrator should be able to take on a case and complete it in a matter of weeks.

Finally, if the matters in dispute and the outcome need to remain confidential, then it would be better to proceed to arbitration which is a private resolution service; whereas cases being litigated are matters of public record. Due to the fact that the parties in dispute are mentioned on Her Majesty's Court Service website, copies of the claim forms may be obtained from the court and the judgments can be obtained electronically or in person at the court.

Independent expert determination

An independent expert is similar to an arbitrator in that it is a person whom is a specialist in the field of the disputed matters, however, the differences make this type of ADR more suitable to some cases than arbitration.

As there is not normally the need to make submissions to an independent expert, unlike an arbitrator, the costs are usually lower. The expert is employed to make their own decision on the findings of facts and law, if appropriate and does not have to stick to a determination within the range made known to them by the parties (unlike arbitration).

Therefore, for a lower value dilapidations or service charge dispute, where the parties have not instructed surveyors or solicitors, an independent expert would be preferred compared to an arbitrator. The parties may leave it to the expert to determine the result of the dispute speedily and without much input from them.

However, there are two disadvantages of an independent expert determining the dispute rather than an arbitrator:

- First, the decision, known as the determination, does not have to be within the range of sums in dispute between the parties. Using the example stated above, on arbitration, with a dilapidations dispute, the tenant or former tenant may wish to offer no higher than say, £80,000, whereas the landlord wishes to obtain £200,000.

  The expert is permitted to determine the outcome below £80,000 or higher than £200,000; so there is a risk to both parties, that does not exist with arbitration, that there can be no guarantees of a minimum sum that a landlord would be able to receive in the determination or a maximum "not to exceed figure" in the case of a tenant.

  Clearly, the reason for the difference between the two roles is that the arbitrator is said to be acting judicially, whereas the independent expert is using their own skill and judgment to determine the dispute and does not have to seek submissions from the parties.

- Second, a party whom is not content with the determination cannot bring court proceedings, unlike an arbitration referred to above. The only route to challenging the determination is to allege professional negligence on the part of the independent expert.

This is an expensive and risky route, as there has never been a successful challenge at the High Court of an expert's determination, albeit there have been many attempts. However, there have been a number of cases that have been settled by the parties (aggrieved party and independent expert) prior to the cases being litigated and the matter entering the public domain.

Early neutral evaluation

Early neutral evaluation is the term given to the process whereby a particular point at issue (not usually the whole dispute) is referred to an appropriately qualified person, whose opinion is likely to be respected by the parties. With the consent of all parties in a dilapidations or service charge dispute, a surveyor, construction engineer or an accountant, if an audit/service charge case, may provide an early neutral evaluation on the difference between the parties.

The Technology and Construction Court, part of the High Court, has an early neutral evaluation scheme whereby one of the judges may determine the issue or issues concerned. Where proceedings are already before that court, this ADR process can still take place and the scheme provides that either a different judge or the judge hearing the case already will determine the differences between the parties. If the judge hearing the case deals with this ADR, then that judge will take no further part in the proceedings once they have produced the evaluation, unless the parties expressly agree otherwise.

Unless the parties agree otherwise, the evaluation decision will not be binding on the parties. The parties have little to lose by referring their "sticking point" to the third party where it is not binding, so it is similar to mediation in its attractiveness as an ADR process.

Stage 4: Litigation

Litigation considerations

If ADR has not produced an agreed settlement, or one or both of the parties have refused to consider it, then the default position for the landlord or the tenant is to litigate.

Following recent updates to the CPR and recent case law, the message has come out loud and clear that courts see litigation as the final stage in a process and not any earlier.

In keeping with the warning contained in the CPR Default Protocol, there have been sanctions on costs for parties whom did not wish to consider alternative routes to settling the case or unreasonably refused ADR possibilities (see the case of *Halsey* v *Milton Keynes General NHS Trust* [2004] EWCA Civ 576.

When a trial has ended, one of the parties may put forward a submission on costs on the basis that the other party unreasonably refused to take part in ADR. The *Halsey* case, which went to the Court of Appeal, identified six factors that may be considered relevant to any such costs submission:

1. The nature of the dispute.
2. The merits of the case brought by the claimant.
3. The extent to which other settlement methods have been attempted.
4. Whether the costs of the ADR would be disproportionately high to the level of the sums in dispute.
5. Whether any delay in the setting up and attending the ADR discussions and meetings would have been prejudicial.
6. Whether the ADR would have had a reasonable prospect of success.

So, the general view is that a party which does not wish to consider the alternatives available to settling the dispute and instead proceeds directly to litigation may suffer a penalty on costs regardless of whether that party is seen as successful at court.

Lawyers have a surprisingly minimal role in most commercial property dilapidations claims. The single most important reason for this fact is that most claims are commenced, negotiated and settled solely by surveyors. This may be appropriate in some cases at the pre-litigation stage, but the landlord, tenant (or their surveyors) should adopt the default position of involving their specialist lawyer during all aspects of the claim.

Dilapidations are no more than damages claims and the lawyer should not be sidelined. Lawyers will, in most cases, justify their involvement with advice on:

- the liability
- negotiation tactics
- potential for litigation
- drafting binding settlement agreements.

Preparing for litigation

Litigation will commence when:

- the landlord claimant forms the view that the dilapidations dispute has become unsettleable via a negotiated route or ADR, etc
- the landlord claimant has completed all necessary disclosure in accordance with legal obligation, in particular compliance with the CPR Default Protocol
- the landlord takes the positive step of issuing court proceedings.

It is not the purpose of this book to deal with all aspects of civil litigation procedure under the CPR 1998. Lawyers experienced in dilapidations claims should always be consulted at the point at which a dispute requires the involvement of the courts.

Adherence to the CPR Default Protocol

The landlord claimant must have complied with both the letter and the spirit of the CPR Default Protocol during the pre-litigation stages. This is a straightforward process and will require the landlord to issue full details of the claim in accordance with the RICS Guidance Note. It will also need the landlord to have commissioned a section 18(1) of the Landlord and Tenant Act 1927 valuation if appropriate and to have made every attempt to settle the dispute without litigation.

Choice of court

Most dilapidations disputes are tried in the Technology and Construction Court where the technical complexity of dilapidations claims is well suited to the expertise of the judges. However, some parties choose to bring disputes into the Chancery Division of the High Court.

Witness statements and evidence

These documents will be required by the landlord at the point that proceedings are commenced. The process of collating, verifying and checking the documentation is generally the responsibility of the lawyer. Statements and evidence required will include:

- the lease
- proof of service
- proof of breaches of lease covenant.

CPR Part 35 governs the role of expert witnesses. In a dilapidations claim this role will be taken by one or all of the following:

- Building surveyor.
- Valuer.
- Quantity surveyor.
- Engineer.

Each expert should pay particular attention to the RICS practice statement and guidance note for *Surveyors acting as Expert Witnesses*. This document mirrors the CPR Part 35 in clarifying that the expert's primary duty is to the court and not the client.

Disclosure

Disclosure of documentation relevant to proceedings is governed by CPR Part 31, which is the detailed procedural rule used by the courts.

Part 31 applies to areas of civil litigation and requires a party to disclose only the documents:

- on which they rely which adversely affect their own case
- which adversely affect another party's case or support another party's case.

Part 31 creates an obligation to "make a reasonable search for documents". This will be judged objectively in the event of an allegation that documents that were in existence should have been supplied.

Summary

The landlord and tenant will be best served if the tactical and procedural issues of litigation have been investigated in detail prior to litigation to ensure that all documentation accords with these obligations.

The potential for any dilapidations dispute to litigate should be borne in mind from the start of the claim, when the schedule of dilapidations is served. The potential consequence of all that is said and done by landlord and tenant surveyors and should be borne in mind from the inception of a dispute.

Expert reports

Each party whom employs surveyors to appear for them at court will have to understand that the surveyor's duty is not to "win at all costs" for their client. Although their fees are paid for by their client, a surveyor's duty is to assist the court in coming to the correct decision on the matters before it.

This also means that a surveyor who had been on a fee agreement based on the percentage of savings made (if acting for an occupier) or based on the amount of the final sum obtained, if acting for a landlord, will have to ensure that the fee basis is moved to a different basis when acting as expert witness, or consider stepping down as expert on the basis that the perception of the impartiality of their evidence is tarnished by self-interest.

The usual fee basis is based on an hourly rate for writing the report and appearing in court, or an hourly rate for the report writing and a daily rate for the court appearances.

14 Common Disputes — Legal Issues

Lease interpretation disputes

A guide to interpretation

Like any process, a dilapidations or service charge claim must start somewhere and the extent and nature of any permissible claim will be dependant to a large extent on the lease terms, covenants and obligations. The implications of the lease terms may not always be readily self-apparent and so before a claim can be made or defended the surveyors engaged will need to carefully interpret the lease.

Sadly, it is also at this early lease interpretation stage in proceedings that so many claims seem to go wrong. This is often because surveyors tend to adopt a mistaken, pedantic and overly literal approach to interpretation. Commonly, this is because the surveyor preparing or defending the claim has traditionally suffered from a lack of guidance or training on this vital aspect and at best has only a limited understanding of the principles of interpretation.

When considering the basic approach to interpretation, it helps to be able to refer back to how the courts approach such matters. To this end, Lord Hoffman gave excellent guidance on interpretation in *Investors Compensation Scheme Ltd* v *West Bromwich Building Society (No 1)* [1998] 1 WLR 896.

Lord Hoffman's principles of interpretation

Lord Hoffman's guidance was somewhat lengthy but because of the importance of this issue, it is set out below in full.

My Lords, ... I think I should preface my explanation of my reasons with some general remarks about the principles by which contractual documents are nowadays construed. I do not think that the fundamental change which has overtaken this branch of the law, particularly as a result of the speeches of Lord Wilberforce in *Prenn* v *Simmonds* [1971] 1 WLR 1381, 1384–1386 and *Reardon Smith Line Ltd* v *Yngvar Hansen-Tangen* [1976] 1 WLR 989, is always sufficiently appreciated. The result has been, subject to one important exception, to assimilate the way in which such documents are interpreted by judges to the common sense principles by which any serious utterance would be interpreted in ordinary life. Almost all the old intellectual baggage of 'legal' interpretation has been discarded. The principles may be summarised as follows:

(1) Interpretation is the ascertainment of the meaning which the document would convey to a reasonable person having all the background knowledge which would reasonably have been available to the parties in the situation in which they were at the time of the contract.

(2) The background was famously referred to by Lord Wilberforce as the 'matrix of fact', but this phrase is, if anything, an understated description of what the background may include. Subject to the requirement that it should have been reasonably available to the parties and to the exception to be mentioned next, it includes absolutely anything which would have affected the way in which the language of the document would have been understood by a reasonable man.

(3) The law excludes from the admissible background the previous negotiations of the parties and their declarations of subjective intent. They are admissible only in an action for rectification. The law makes this distinction for reasons of practical policy and, in this respect only, legal interpretation differs from the way we would interpret utterances in ordinary life. The boundaries of this exception are in some respects unclear. But this is not the occasion on which to explore them.

(4) The meaning which a document (or any other utterance) would convey to a reasonable man is not the same thing as the meaning of its words. The meaning of words is a matter of dictionaries and grammars; the meaning of the document is what the parties using those words against the relevant background would reasonably have been understood to mean. The background may not merely enable the reasonable man to choose between the possible meanings of words which are ambiguous but even (as occasionally happens in ordinary life) to conclude that the parties must, for whatever reason, have used the wrong words or syntax. (see *Mannai Investments Co Ltd* v *Eagle Star Life Assurance Co Ltd* [1997] 2 WLR 945)

(5) The 'rule' that words should be given their 'natural and ordinary meaning' reflects the common sense proposition that we do not easily accept that people have made linguistic mistakes, particularly in formal documents. On the other hand, if one would nevertheless conclude from the background that something must have gone wrong with the language, the law does not require judges to attribute to the parties an intention which they plainly could not have had. Lord Diplock made this point more vigorously when he said in *The Antaios Compania Neviera SA* v *Salen Rederierna AB* [1985] AC 191, 201:

> '... if detailed semantic and syntactical analysis of words in a commercial contract is going to lead to a conclusion that flouts business commonsense, it must be made to yield to business commonsense.'

Obviously, lease phrasing, circumstances, background and the matrix of the facts will vary from case to case. Ideally, however, if surveyors approach their lease review and subsequent claim or defence services from a common sense "reasonable man" starting point, then they are more likely to avoid disputes and resolve cases fairly and in a time effective manner than would otherwise be the case.

Interpretation dispute resolution

Where surveyors find themselves in entrenched positions over the interpretation over a single lease clause, they should also take care not to let any difference of professional opinion distort the claim handling and resolution negotiations in a disproportionate manner. If the claim value consequence of any disputed interpretation item is relatively minor, then pragmatism and common sense should in most cases be allowed to win the day.

However, if the interpretation dispute between surveyors concerns an item of claim of high value or of significant consequence for the viability of the claim, then the taking of an independent expert third party opinion will often be the quickest and most appropriate way to simply and effectively resolve the dispute. On these high value/significance interpretation dispute occasions it is worth the disputing surveyors agreeing to take a single counsel opinion from a suitably experienced and qualified joint appointed counsel/QC.

The consequences of both sides belligerently persisting with disparate lease interpretations is that one side or another will ultimately lose their argument in court (perhaps wrongly) and the opportunity to avoid suffering significant (or perhaps catastrophic) litigation costs will have been missed for what can only be put down to surveyor vanity or pride.

Statutory compliance disputes

Lease and statutory obligations

In most modern institutional leases, there will be a tenant's covenant and obligation concerning compliance with acts of Parliament and subservient legislation (regulations, byelaws, orders, etc). Such clauses will typically also require the tenant to comply with the requirements of competent authorities such as local authority regulatory departments, The Health and Safety Executive or other empowered bodies such as CORGI, etc.

In most "compliance" clauses there will be an express obligation on the tenant to actively comply with statute and regulations but even where this clause requirement is absent from the lease, the obligation on the tenant will exist due to the nature of the statutory provisions. Most compliance clauses will also contain a further obligation on the tenant to "indemnify" the landlord against any claims arising from a want or neglect of compliance.

Dilapidations claims in the context of a tenant's breaches of the compliance obligations of the lease remain a greatly misunderstood area of dilapidations claims. There continues to be debate within the professional community whether or not the intention of compliance clauses is to provide a "protection" or a "benefit" to the landlord. The reality, however, is normally far simpler and in most cases the suggestion of creating a benefit for the landlord is mistaken.

Where a compliance clause includes a general tenant indemnity obligation in the landlord's favour, this is clearly a protective measure on behalf of the landlord. Quite often, landlords will be the first recipients of enforcement action from appropriate authorities. For example, should a property become so dilapidated that it is found to be unsafe then more often than not the landlord will be the recipient of the local authority's dangerous structures notice and enforcement action in the first instance. This in turn would give rise to dilapidations claims and action by the landlord upon the tenant under the lease where the landlord will be protected from suffering costs due to the tenant's general indemnity undertakings.

However, the argument suggesting that a compliance clause may provide a "benefit" to the landlord appears based on the fact that a tenant may on occasion need to undertake works to comply with statute or regulations, etc that require an improvement to the property at the tenant's expense but to the landlord's long-term benefit.

Compliance required improvement works

The duty to undertake improvement works are often contested by tenants on the basis that the improvement works will exceed the "repair" works obligations of the lease. For some reason, tenants who seek to contest compliance required improvement works mistakenly believe that a private contractual agreement between two parties concerning "repair" liabilities will take preference over their statutory compliance obligations.

To illustrate the failings of this "benefit" defence argument one only needs to ask oneself whether a tenant within a property with say, 10 years remaining on the lease and a requirement to undertake an improvement in order to comply with a new statutory obligation, would refuse to undertake the improvement because it exceeded their repair obligations; and would instead abandon the property and take a separate but concurrent second lease on an alternative, already compliant property as the most appropriate means of meeting their statutory compliance obligations. It is common sense that no tenant would act in such a manner.

All reasonable tenant-like users with statutory obligations that require improvements within a property should undertake improvement works as necessary to comply with the law, regardless of whether or not they exceed a repair covenant obligation. Many tenants may see this obligation as being unfair but it need not be the case. Parliament has already considered the issue of unfair benefit to landlords arising from improvements necessary to comply with statute.

Under Part I of the Landlord and Tenant Act 1927 as amended by Part III of the Landlord and Tenant Act 1954, there are already statutory procedures in place where a tenant in such circumstances can serve notice on the landlord of their intentions to undertake improvements to the property in order to comply with statute (which the landlord cannot reasonably refuse).

Subject to complying with the statutory procedures contained in the Acts and serving notices at express points during the life of the lease, the tenant can claim "compensation" at the end of the lease term for the value of any improvements made to the landlord's property by the tenant (or their predecessor in title). Any compensation for improvements will then be offset against any end of term dilapidations claim.

As there is an existing and adequate process for a tenant to claim compensation for improvements, there is no "benefit" to the landlord and no legitimate reason for the tenant not to undertake the improvements.

Moreover, and from a dilapidations perspective, if the tenant fails to undertake the compliance required improvements, no legitimate grounds for the landlord to claim damages arise as the lost benefit to the landlord is offset by the equal level of compensation saved that would have been due to the tenant.

Non-improvement compliance works

Where the tenant fails to undertake "compliance" required works of testing, repair or maintenance of existing features or new additions/ alterations, there will generally continue to be reasonable grounds to claim damages for the breach (regardless of whether or not the tenant has any repairing obligations under the lease).

If for example the tenant is unable to demonstrate that they have regularly tested and maintained the electrical installations of the property in accordance with statutory and regulatory requirements, then there will be an actionable breach of the compliance clause. The remedy to be sought for the breach would be for the tenant to undertake the necessary testing. It would be reasonable for the landlord to seek copies of test certificates and reports once the tenant has remedied the breach so they may satisfy themselves that the breach has indeed been remedied.

Should the test reports identify further wants of repair, then the landlord can at that stage serve further notices of breaches of the repair covenants, requiring the tenant to carry out those wants of repair.

Claim discounts for "betterment"

It is generally accepted that a claim for dilapidations should not include any element of improvements above the standard that was passed to the tenant on the first day of the lease. However, this can cause both parties a problem if the original standard of fitting or equipment present at lease grant is no longer manufactured.

An example of this is the provision of black and white CCTV at a building that is let to a single tenant, whom is responsible for the repair and maintenance of that system. If that CCTV equipment is in disrepair at the end of the lease, it is only right that the landlord should request the replacement of it in the schedule of dilapidations.

However, the tenant's advisors (surveyors and lawyers) could raise the point that the tenant only has to pay towards a black and white system in relation to this item on the schedule of dilapidations; a colour system would be an improvement. If it is generally accepted that an improvement is the giving back to the landlord of something better than was granted by the landlord, then the landlord should have to concede this point.

The problem then arises that it would not be possible to purchase a replacement black and white CCTV system, as technology has moved on and they are superseded by colour systems.

One way of dealing with what could easily become a stalemate position, is a "sharing arrangement" between the parties: the tenant agrees to pay a sum equivalent to what a black and white CCTV system would cost and the landlord picks up the difference between that and a colour system. Therefore, the tenant is content that they are not paying for an improvement to the landlord's building; and the landlord is content that they are recovering a sum that would have been applicable if a black and white system had been sourced. The courts consider this sensible approach as being that claim "survives" in principle for the purposes of section 18(1) of the Landlord and Tenant Act 1927 but that a "discount" for betterment is to be applied to the damages sought to reflect the betterment gained.

"Tenant-like" user disputes

It has been held by the courts that there is an implied obligation on a tenant under a lease to use and occupy a demised property in a "husband-like" or "tenant-like" manner. This implied obligation arose from historic judgments concerning claims against residential tenants on "year-to-year" lease terms and was famously summarised by Denning LJ in *Warren* v *Keen* [1954] 1 QB 15 as being an implied obligation:

> The tenant must take proper care of the place. He must, if he is going away for the winter, turn off the water and empty the boiler. He must

> clean the chimneys, when necessary, and also the windows. He must mend the electric light when it fuses. He must unstop the sink when it is blocked by his waste. In short, he must do the little jobs about the place which a reasonable tenant would do. In addition, he must, of course, not damage the house, wilfully or negligently; and he must see that his family and guests do not damage it: and if they do, he must repair it. But apart from such things, if the house falls into disrepair through fair wear and tear or lapse of time, or for any reason not caused by him, then the tenant is not liable to repair it.

However, while the often quoted case of *Warren* v *Keen* is a useful starting point for considering the implied obligations of a tenant-like user, it concerned a claim on the equivalent of a modern Assured shorthold tenant and so is not entirely appropriate for wider application in commercial tenancies. It has also been observed that it has yet to be tested in a more modern court and is possibly somewhat inadequate in modern times when considering the "matrix of facts" and modern tenant statute imposed obligations.

In the 54 years that have passed since the *Warren* v *Keen* judgment, there have been substantial statutory and regulatory developments that have material and significant implications for determining what a tenant-like user (with no express contractual/lease repair obligation) would or would not be allowed to do in a property. Also, when the courts are asked to consider the interpretation of a contractual term or obligation, the courts will normally seek to do so in a way that would not of itself result in the parties being in breach of the law. Consequently, the courts will seek to take into consideration not only contractual obligations but also materially relevant statutory and regulatory obligations.

For example, a tenant "occupier" of demised premises will have a duty of care to visitors (and even trespassers) under the Occupiers' Liability Acts 1957 and 1984. If the state of the property poses a hazard or risk to visitors; and where the landlord has no repair obligations that the tenant can ask the landlord to perform, then there is a good case to argue that the tenant, being a good tenant-like user "occupier" would undertake those repairs in order to manage their risk and liability.

By way of further example, there is also a common misconception that for commercial properties, there is no statutory tenant-like duty to repair a premises in the absence of any express lease obligation. However, commercial tenants who are subject to the Health and Safety at Work Act 1974 are also subject to Regulation 5 of the

Workplace (Health, Safety and Welfare) Regulations 1992 that states (with added emphasis):

> The workplace and the equipment, devices and systems to which this regulation applies *shall be maintained* (including cleaned as appropriate) in an efficient state, in efficient working order and *in good repair*.

Again, where the landlord has no repair obligations that the tenant can ask the landlord to perform, then there is a good case to argue that the tenant, being a good tenant-like user "employer" must undertake the necessary maintenance and repairs to honour their statutory obligations.

In short, the tenant-like user of 1954 arguably no longer exists and a reliance on the judgment of *Warren* v *Keen* [1954] 1 QB 15, without considering the modern statute-imposed user obligations for any particular class and type of tenant, is no longer appropriate.

Where disputes arise relating to tenant-like user obligations, a surveyor must therefore have regard to the full "matrix of facts" of modern tenant-like user, employer and occupier legislation when considering what implied or otherwise imposed obligations the tenant may have for repairs and maintenance at the property.

Time barred claims: The Limitation Act 1980

The purpose behind this Act, which replaced an earlier "Statute of Limitations", was to prevent "stale" claims being brought to court after a certain period of time. The intention behind the Act is to set out a series of dates where a party wishing to bring an action under contract or in tort had to bring the case to court within a set period of time, otherwise they would lose the right to bring that claim, ie they were "time barred".

The Act details the timings — in relation to dilapidations and service charge claims, these are either six or 12 years.

Therefore, from the date that a contractual breach takes place or a right to bring a claim in court exists, the claimant has a period of time before which they must present a claim form at either the county court or the High Court. If that party is even one day too late, based on the time set out in the Limitation Act 1980, then that claim cannot be accepted by that court.

In relation to service charge claims, as these are a debt and can never be anything else, ie they are not damages claims or claims under negligence, then the period of time is six years from when the sum concerned was due. In relation to dilapidations claims, these are either six years where a lease is simply signed by the parties or 12 years where the lease is under seal. As most leases of large premises with large companies are under seal and not simple signatures, then the usual term is 12 years.

Practitioners need to be aware of the implications of Limitation Act 1980 as some issues go on for two or three years or more and it is only when a claim form has been issued at the county court or the High Court that the time period is ended. Therefore, practitioners need to be aware that even though detailed negotiations may be underway, their client's claim will remain subject to the six or 12-year limitation period. It has been known that some practitioners will continue debates over a longer and extended period, knowing that the limitation period is about to expire and therefore the sums concerned could not be recovered, in any case, by way of court action. Clearly, if court action cannot take place, then neither can mediation, arbitration, independent expert referral or any of the other dispute resolutions services available to deal with any disputes.

For example, why would a party wish to take a case to mediation or to an independent expert or arbitration if that same case cannot be taken to litigation?

There have been a number of cases before the High Court and or the professional bodies such as the Bar Council (whom regulate barristers), the Law Society (now the Solicitors Regulation Authority) or the Royal Institution of Chartered Surveyors (RICS) relating to practitioners who have not advised of the implications of the Limitation Act 1980 and the cases that they have been instructed to deal with have had to end at that point and not been able to be referred to court or a third party.

A practitioner in a service charge or dilapidations dispute ignores the implications of the Limitation Act 1980 at their peril.

Prejudicial conflicts of interest

The independence and reliability of the surveyor should not be prejudiced by the surveyor engaging in, what could be viewed as, overly partisan negotiations or conduct.

The need to adopt an independent or an impartial position may also, for example, potentially compromise other pre-existing business relationships where the business cost of a client/consultant relationship souring exerts undue pressure on the way the claim is to be conducted.

The handling of dilapidations or service charge claims and the conduct of the parties is unlikely to be conducted fairly or in an independent and unbiased manner where a conflict of interest exists or could be considered to exist. Such conflicts will invariably lead to disputes.

A surveyor may have a conflict of interest that is not always immediately apparent, however, if one exists or is allowed to arise then it may have significant consequences for the prospects of success of a claim or defence. Conflicts of interest whether recognised or not, will typically hamper claim resolution proceedings and will give rise to disputes between the surveyors.

As all surveyors acting on dilapidations or service charge claims perform an expert function (see Chapter 4), the guidance given by the courts to experts over conflicts of interest equally applies to surveyors attending to the early stages of a claim.

In *Armchair Passenger Transport Ltd* v *Helical Bar plc* [2003] EWHC 367 (QB), Nelson J stated the following conflict of interest principles:

(i) It is always desirable that an expert should have no actual or apparent interest in the outcome of the proceedings.

(ii) The existence of such an interest, whether as an employee of one of the parties or otherwise, does not automatically render the evidence of the proposed expert inadmissible. It is the nature and extent of the interest or connection which matters, not the mere fact of the interest or connection.

(iii) Where the expert has an interest of one kind or another in the outcome of the case, the question of whether he should be permitted to give evidence should be determined as soon as possible in the course of case management.

(iv) The decision as to whether an expert should be permitted to give evidence in such circumstances is a matter of fact and degree. The test of apparent bias is not relevant to the question of whether or not an expert witness should be permitted to give evidence.

(v) The questions which have to be determined are whether (i) the person has relevant expertise and (ii) he or she is aware of their primary duty to the Court if they give expert evidence, and willing and able, despite the interest or connection with the litigation or a party thereto, to carry out that duty.

(vi) The Judge will have to weigh the alternative choices open if the expert's evidence is excluded, having regard to the overriding objective of the Civil Procedure Rules.

(vii) If the expert has an interest which is not sufficient to preclude him from giving evidence the interest may nevertheless affect the weight of his evidence.

Furthermore, in the case of *Toth* v *Jarman* [2006] EWCA Civ 1028, the Court of Appeal judges Arden J, DBE and Wall J collectively held that:

119. (a) conflict of interest could be of any kind, including a financial interest, a personal connection, or an obligation, for example, as a member or officer of some other body. But ultimately, the question of what conflicts of interest fall within this description is a question for the court, taking into account all the circumstances of the case.

120. Without wishing to be over-prescriptive ..., we are of the view that consideration should be given to requiring an expert to make a statement at the end of his report on the following lines:

(i) that he has no conflict of interest of any kind, other than any which he has disclosed in his report;
(ii) that he does not consider that any interest which he has disclosed affects his suitability as an expert witness on any issue on which he has given evidence;
(iii) that he will advise the party by whom he is instructed if, between the date of his report and the trial, there is any change in circumstances which affects his answers ...

121. As we see it, a form of declaration to this effect should assist in reminding both the expert and the party calling him of the need to inform the other parties and the court of any possible conflict of interest.

Given the expert's duties with regards to conflicts of interest expected by the court, it is clearly not in a client's interest to appoint a surveyor at the start of the claim who is incapable of acting on their behalf as an "expert" at a later stage should the dispute eventually be referred to the courts for resolution.

Before accepting an appointment, the surveyor should therefore consider the issue of conflicts of interest very carefully and ethically ought to protect their client and decline the appointment where a conflict exists that could be prejudicial to the client's claim or defence prospects. Ethically, the surveyor should seek to protect their client from the difficulties that such an appointment would give rise to and so should not quote for dilapidations service on a performance-related fee basis.

For example, a conflict of interest could also be said to exist where there is a long-standing client/consultant or agent relationship between the client and the surveyor's practice, separate to the dilapidations matter. This is because the relationship could be at risk or damaged if the surveyor fails to structure a claim or obtain a settlement that does not meet the client's possibly inflated or misguided expectations. In other words, the existence of a relationship between claimant and advisor may be considered to exert an undue influence that (perhaps unwittingly) prejudices the conduct and actions of the parties.

During claims handling proceedings, if either side has any concerns over the independence of the other's appointed surveyor(s) and/or experts and the possible existence of a conflict of interest, then the surveyors should be invited to make similar declarations to those declarations of "experts" expected by the courts (see above).

As it is good practice and expected by the court for a surveyor and/or expert to disclose potential conflicts of interest at the earliest possible point, the authors suggest that it is now appropriate and professionally appropriate good practice for suitable declarations to be made by the surveyors in their first open claim documents or defence responses.

Misrepresentations: Innocent, negligent or fraudulent?

When attending to dilapidations claims, both the landlord and tenant parties will make various representations, statements and claims that set out their opinion of the dilapidations issues and any associated damages claim. These claims should be truthful and properly supported, otherwise the party making rash or deceitful representations and claims will suffer potentially very serious consequences should the claim go to court.

Honesty and good faith

A fairly obvious factor that causes difficulties and promotes dispute in many dilapidations and service charge claims proceedings is the perception of dishonest conduct. Given the nature and often sizeable cost consequences of such claims, it is not unreasonable to suggest that any costs or damages claims and negotiations are procedures that should be regarded as *uberrimae fidei*, eg that they must be conducted in utmost good faith.

In *Blisset* v *Daniel* (1853) 10 Hare 493, it was held that the duty of good faith which subsists between partners (or parties) extends to any process, whether by negotiation or otherwise, designed to bring about the termination of that relationship.

Where there are concerns over the truthfulness of an aspect of the claim or defence negotiations, it may help resolve the situation if the suspect parties are reminded of guidance of Millett J in *Logicrose Ltd* v *Southend United Football Club Ltd* [1988] EGCS 114 that parties to negotiations "do not owe each other a duty to act reasonably, but only to act honestly".

If both surveyors are in principle endeavouring to remain fair and reasonable in their dispute resolution dealings and are seeking a fair settlement; then both surveyors should be prepared to share and openly disclose any genuine and materially relevant evidence that they seek to rely upon.

For example, unfortunately it is not uncommon to find that surveyors may "harmlessly" invent or distort third party remedial works cost "evidence" that they claim is gained from another similar project and that they then refer to in support of an otherwise unsupported claim or defence position. This is often because the surveyor possibly lacks the experience to appreciate that what they are doing is potentially a serious misrepresentation and the surveyor may never envisage having to disclose the fabricated cost information.

Where there are concerns over representations of undisclosed third party cost evidence, then it is reasonable to request copies of the evidence to be made available. On occasion, the surveyor who made the cost claims then can't provide the alleged evidence when requested to do so. More often than not, even if they are able to disclose any past tender figures, close scrutiny will commonly show that the figures are of little material relevance to the current claim.

Where rash, reckless or deceitful claims or defences are found to have been made, the surveyor that has made the dishonest representations will struggle to maintain credibility. The adoption of a claim or defence strategy based on deceit is abhorrent to most professional surveyors and the RICS. It goes against the core codes of conduct for chartered surveyors and provides ground for disciplinary proceedings to be commenced against the suspect surveyor and/or their firm.

Deceit and dishonesty in claims or defence proceedings also has serious and potentially criminal prosecution ramifications for both misrepresentation and fraud (see below).

Understanding representations

Representations

The Oxford English Dictionary defines "representation" as being:

> noun:
>
> 1. the action or an instance of representing or being represented;
> 2. an image, model, or other depiction of something;
> 3. (representations) statements made to an authority to communicate an opinion or register a protest.

In a professional context, a representation could be said to be a statement made by one party to another; with regard to some existing fact or to some past event; which is of material relevance.

Misrepresentation

A misrepresentation is an untrue statement of fact, made by one party to another; either before or at the time of making; with the intention that the person to whom the statement is made shall act upon such misrepresentation, and they do so act.

Any misrepresentations made during the course of proceedings may result in actions in relation to the misrepresentations under the law of tort (for deceit) or under the Misrepresentation Act 1967. The severity of any action or consequence will ultimately depend on the misrepresentation and whether it was innocent, negligent or fraudulent in nature.

Misrepresentations can also be passive in nature. In *Ross River Ltd* v *Cambridge City Football Club Ltd* [2007] EWHC 2115 (Ch), Briggs J considered that although silence as to material facts is not in general capable of constituting a misrepresentation, it may do so where a party to the negotiation/contract is under a positive duty of disclosure or where an existing relationship between the parties imposes an obligation of disclosure.

Innocent misrepresentations

Innocent misrepresentations are made where the maker of the statement has reasonable grounds for believing in its truth.

Negligent misrepresentations

A negligent misrepresentation is a statement of "fact" made by a party who had reasonable grounds to believe; and they did believe that the facts represented were true.

A negligent misrepresentation cannot be fraudulent, provided the party had an honest belief in the truth of the statement made.

When a misrepresentation claim is based in negligence, the Misrepresentation Act 1967 creates a reverse onus on the defendant to disprove negligence.

Fraudulent misrepresentations

The concept of a fraudulent misrepresentation was set down in the judgment of *Derry* v *Peak* (1889) 14 App Cas 337 when it was held that:

> A Fraudulent Misrepresentation is a false statement made knowingly or deliberately or without belief in truth — or recklessly without caring whether it was true or not.

In the case of *Thomas Witter Ltd* v *TBP Industries Ltd* [1996] 2 All ER 573, it was stated that for a misrepresentation to be considered to be a fraudulent misrepresentation:

> it must be made with the intention that it should be acted on and it is in fact acted upon. *Male fides* are not a prerequisite for a fraudulent misrepresentation to be proved ... Recklessness is only evidence of fraud — not proof, unless it amounts to a flagrant disregard for the truth and so is also dishonest.

In *Doyle* v *Olby (Ironmongers) Ltd* [1969] 2 QB 158, Lord Denning held that:

> Fraudulent Misrepresentations are a specific tort of deceit. The defendant is bound to make reparation for all the damage flowing from the fraudulent inducement

In *Ross River Ltd* v *Cambridge City Football Club Ltd* [2007] EWHC 2115 (Ch), Briggs J also held that in a case where fraudulent material misrepresentations have been deliberately made, with a view to improperly influence the outcome of negotiations in favour of the maker and his principal by an "experienced player" in the relevant market/field, then "there is a powerful inference that the fraudsman achieved or endeavoured to achieve his objective, at least to the limited extent required by the law, if the his fraudulent improper influence was actively in the mind of the recipient when the contract or negotiated agreement was considered or came to be made".

Briggs J also held that "an analysis of misrepresentation, particularly in relation to materiality and inducement, requires a comparison to be carried out between the statement actually made, and the truth, rather than between the statement made and silence, ie no statement."

Fraud and the Fraud Act 2006

The Fraud Act 2006 came into force in January 2007 and has amended those parts of the Theft Act 1968 relating to dishonesty and fraudulent statements that have been made in relation to recovering monies from one party against another.

The Act is described as being:

> An Act to make provision for, and in connection with, criminal liability for fraud and obtaining services dishonestly.

A number of practitioners in service charge or dilapidations disputes have been concerned about the implications of this Act, as the Act does not change any of the procedures that were in place before the Act came into force in January 2007 in relation to dishonest or fraudulent statements.

The only change between the Theft Act 1968 offences and the Fraud Act 2006 offences is that the position is worse for the "offender" under the Fraud Act 2006. Previously, a party had to actually suffer a loss of money or other measurable quantity prior to being able to alert the Police and the Crown Prosecution Service. If that loss was not suffered, then the matter would not be investigated.

However, under the Fraud Act 2006, that party does not actually have to suffer a loss; they only have to be placed in a position where they may have suffered a loss.

It would be useful at this stage to define the definition of a "false representation" under that Act.

Section 2 of the Fraud Act 2006 states:

> Fraud by false representation
> (1) A person is in breach of this section if he —
> (a) dishonestly makes a false representation, and
> (b) intends, by making the representation —
> (i) to make a gain for himself or another, or
> (ii) to cause loss to another or to expose another to a risk of loss.

Section 2(2) defines a false representation as one that is untrue or misleading, and the person making it knows that it is, or might be, untrue or misleading.

Punishment for fraud

As the Act is an act that sets out criminal offences, section 1(3) states that a person who is guilty of fraud is liable on summary conviction, to imprisonment for a term not exceeding 12 months or to a fine not exceeding the statutory maximum (or to both); or on conviction on indictment, to imprisonment for a term not exceeding 10 years or to a fine (or to both).

Fraud in practice

As stated above, the duties of a practitioner in relation to seeking monies from another party have not changed, albeit general commentary that some of the press and discussion groups may have stated to the contrary.

Where a surveyor is expressing their opinion on a point, this cannot be in breach of section 2 of the Act due to subsection 3 stating that "representation" means, in this case, any representation as to fact or to the law.

Therefore, the Fraud Act 2006 could only apply when a surveyor is giving a statement as to a factual point. An example of this is where a question has been made to a surveyor: "How many contractors were asked to quote for the work set out in the schedule of dilapidations?" If the answer is given that three contractors were asked to quote, when the fact is only one was asked, then that may be, *prima facie*, a breach of section 2.

Another example would be where a surveyor is seeking to recover VAT on the costs on works in a schedule of dilapidations, where VAT is not recoverable (see elsewhere in this book for reasons for this). The fact that VAT is stated as being applicable is not necessarily a breach of the Act; it could have been included by error on the part of the surveyor or the client. However, where a response to a query on this is made along the lines of: "We have checked the position with our client's accountants and they confirm it to be ..." If that statement (representation) is untrue, then there may be a breach of section 2.

A number of interesting points are worth bearing in mind in relation to fraud including the Fraud Act 2006 which built on earlier criminal offences including the Theft Act 1968.

A party cannot commit fraud by "mistake". The word "representation" is used in the Act and has been used previously under the Theft Act 1968 in relation to the fact that it may not necessarily be a statement. The representation in relation to just the fact of serving a schedule of dilapidations with a series of alleged breaches of contract in relation to the lease itself satisfies the test of the representation which is also clearly a statement in itself. The only way that fraud can be committed in relation to service charge and dilapidations is where a party makes an intentional false statement knowing this it is dishonest in relation to queries being asked, eg where only one quote is obtained. The answer is given that three competitive quotes were obtained and the cheapest was chosen. The reason for this is that it must be a statement of fact rather than a statement of opinion.

The fact that a false statement has been made under the Fraud Act 2006 does not automatically entitle a wronged party to compensation damages. They must sue for tort of deceit in the civil courts as has been established over the last 120 years and, more recently, at the High Court in the case of *Cheshire Building Society* v *Dunlop Haywards (DHL) Ltd* [2008] EWHC 51 (Comm).

In summary, the Fraud Act 2006 has highlighted to a number of practitioners the problems that may come about in relation to false statements. However, this has been overstated somewhat as the Act only applies to statements of fact rather than statements of opinion.

For an offence to have been committed, it must be a factual statement, rather than an opinion and must be of a nature that satisfies the common law test for fraudulent misrepresentations (see above).

15 Common Disputes — Technical Issues

Diminution disputes

Section 18(1) of the Landlord and Tenant Act 1927 and the common law diminution valuations issues are probably the most misunderstood area of dilapidations. To understand why, one must first understand the origins and objective of section 18(1).

From around the 1850s, dilapidations common law developed in a way that generally favoured the cost of the works as being the measure of contractual damages. The courts stopped short of declaring this measure an absolute rule, having initially expressed that they were minded to do so. By the 1920s, this preference was beginning to go into decline, owing in part to the courts' view that some landlord claims were verging on blatant exploitation.

By incorporating section 18(1), Parliament sought to curtail this abuse of position. In the words of Rt Hon Major Owen MP, who spoke in the House of Commons during the passing of the Bill, a landlord's claim for dilapidations would be "restricted to the actual loss" suffered.

Section 18(1) can be viewed in its historical context as a well-meaning and fair-minded attempt to create a more equitable system of governing relations between commercial landlords and tenants. The subsequent effect upon dilapidations claims has been significant. Unfortunately, despite the passage of over 80 years since the introduction of the Landlord and Tenant Act 1927, the interpretation and application of "caps" to claims for damages available under section 18(1) of the Act remain a source of much debate and are often at the heart of dilapidations cases which proceed to litigation.

Supersession and survival concepts

A common dilapidations dispute that arises between surveyors concerns the issue caps to damages available under the second limb of section 18(1) of the Landlord and Tenant Act 1927. In almost all instances, this dispute will arise where the tenant's surveyor may seek to argue (rightly or wrongly) a reduction to the landlord's damages claim, on the basis that some (or all) of the dilapidations remedial works would be "rendered valueless" by alternative works intentions. Surveyors in general have adopted the term "supersession" as a quick reference phrase to such "rendered valueless" circumstances.

Unfortunately, when a surveyor claims supersession, they often rashly assume that the entire disputed item would automatically be rendered valueless because of the end result of any alternative works intentions. However, this is not necessarily the case and surveyors dealing with supersession disputes often neglect to consider "surviving" works properly.

The concept of "survival" relates to claims for works that may be common with both the potential dilapidations remedial works and any alternative refurbishment works that are planned; in other words, aspects of two works intentions that are common if the works are undertaken in sequence and where expenditure incurred in undertaking the common elements would not be rendered valueless (and would therefore "survive" in the dilapidations damages claim).

The concepts of supersession and survival can be further illustrated in the graphic below:

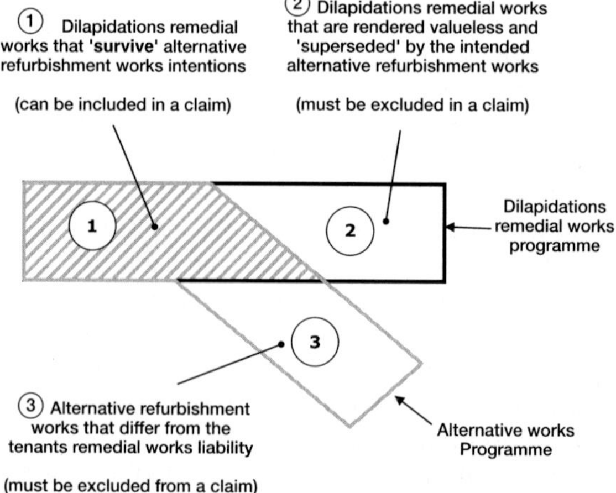

For example, if a landlord had reasonable and legitimate grounds to claim for the stripping-up and replacement of a worn out carpet; but had intentions at the lease end term to strip up the carpet and lay and fit vinyl sheet flooring, regardless of the repair and condition of the carpet; then the stripping-up and disposal works and costs incurred would be common and would "survive", but any claim for the cost of a new carpet being supplied and fitted would be "superseded".

Section 18(1) diminution disputes

"Damage to the reversion"

Numerous judgments have commented on the meaning of the word "reversion" and its relevance to property valuations and lease end dilapidations. The Leasehold Advisory Service has provided a succinct definition:

> A Landlord's interest comprises, basically, the right to receive the rent and the right to have the property back at the end of the lease (this is known as the reversion, because the property reverts to the Landlord's ownership).

The term "reversionary value" is generally accepted as being the value of the interest that the landlord would receive at the end of the lease if the landlord were to sell their interest on the open market. It is the net (residual) sum after deducting all costs and fees, assuming the existence of a willing and reasonable hypothetical purchaser.

The process of valuing the extent by which the reversionary value has been diminished following a breach is known as a diminution valuation. Valuation surveyors who deal with dilapidations will argue that valuation is not an exact science but an art. Valuers often face a lack of relevant evidence and must therefore rely upon their professional judgment, mixed with instinct based upon experience and knowledge of the market. Valuers' opinions may also be influenced — or tempered — by a building surveyor's technical consideration and opinions.

In an ideal world, the key influencing factors for a dilapidations valuer will be similar local and recent transactions that have completed. Commonly though, close comparisons are impossible and so-called "comparable evidence" is merely a guide and should not be deemed to be definitive.

Since no two properties are identical and diminution valuations are rarely straightforward, it is not surprising that surveyors, lawyers and the courts have sought a standardised process that can be adopted in the majority of lease claims. In recent years, dilapidations surveyors have often adopted the diminution valuation approach appended to *Shortlands Investments Ltd* v *Cargill plc* [1995] 1 EGLR 51.

The Shortlands *v* Cargill *approach*

The *Shortlands* approach of appraising the value of the diminution of the reversionary interest relies upon the preparation of two valuations. It is a clear process and dilapidations surveyors and lawyers should be familiar with the ruling. This method of appraisal requires an initial valuation (valuation A) to assess the open market value of the property at lease end with no breaches of the lease. A second valuation (valuation B) seeks to appraise the value of the property in the open market with allowances being made for tenant breaches that would survive a hypothetical purchaser's future intentions for the property. The minimum difference between valuation A and valuation B is said to be the "diminution in value" of the reversion attributable to the tenant's breaches.

In theory, this seems straightforward. In reality, valuation issues on commercial property are often difficult to assess and surveyors seeking to adopt the Shortlands approach need to understand the risks involved. This apparent lack of detailed comprehension frequently leads to dilapidations surveyors, valuers and lawyers adopting overly formulaic and inflexible or irrelevant valuation positions on some claims.

Weaknesses with the Shortlands *approach*

The calculation used in the approach considers the issue from the hypothetical purchaser's viewpoint rather than that of the known vendor (the landlord/claimant).

The valuations start with an estimate of the capital value for the property based upon local market rents and yields. They then make assumptions on the deductions that the hypothetical purchaser would conceivably make for: acquisition expenses; refurbishment works and fees; surviving dilapidations (in valuation B); professional project-related fees; holding costs; and post-works marketing and re-letting costs and fees.

The resulting valuation figure (the purchaser's offer figure) is often claimed to represent the reversionary value for the specific valuation scenario under consideration. However, that figure does not accurately represent the vendor's net reversionary value for the property.

The standard Shortlands valuation approach does not consider all the unavoidable costs and deductions that would be incurred by the vendor. These include: marketing costs; agent and solicitor costs; and tax implications, such as corporation tax or capital gains tax liabilities. Vendor costs and deductions can have a significant and material bearing on the valuation and should not be omitted if the net reversionary value is to be properly appraised.

It is often asserted that the only information that changes between valuations A and B in a Shortlands-style valuation is that relating to the cost of the "surviving" dilapidations works (normally £0 in valuation A and the value of the parts of the dilapidations claim that "survive" supersession review in valuation B). However, does this approach stand up to close scrutiny?

The combined cost of the refurbishment works and the surviving works in valuation B may mean that the works are more complex. They may require or necessitate the involvement of more specialist consultants or statutory consent(s) than would be the case if simpler refurbishment works were to be undertaken in isolation, as allowed for in valuation A. Higher fees and costs will clearly have an effect upon the valuation calculation. The overly hypothetical nature of the Shortlands-style calculation is potentially dangerous. The risk is further highlighted in the not improbable circumstances where the works might influence the reasonable hypothetical purchaser's consideration to acquire the property. Marginally different market rents and/or investment yields may result between the two valuations.

Moreover, the increased value of the total valuation B works and the potentially longer project duration of such works may influence the terms, costs and availability of project funding. This in turn will invariably affect the hypothetical purchaser's allowances and deductions.

The different circumstances of the two valuation scenarios can affect more than the works value. Thus, each separate valuation should be carefully appraised from first principles and the various aspects should be impartially assessed.

In *Simmons* v *Dresden* [2004] EWHC 993 (TCC), HH Judge Richard Seymour QC stated that the Shortlands approach was:

> based upon the assumption that a purchaser of a building in disrepair will either incur expenditure in dealing with the disrepair, or, if he

> accepts the building in the condition in which it in fact is, will modify the price that he is prepared to pay to reflect the fact that the building is out of repair.

The judge highlighted that the "critical weakness" of the Shortlands method was that the calculation was based upon: (i) an assumption of diminution; and (ii):

> the calculation is not designed to test whether there has actually been a diminution in the value of the reversion, but on the assumption that there has been, to calculate what it was.

In other words, the routine adoption of the Shortlands approach often puts the cart before the horse.

Surveyors often contend that the Shortlands approach quantifies the "true loss", although the common law principles of loss support this assertion only in limited circumstances.

Common Law "diminution" disputes

In many respects, the statutory requirements of section 18(1) of the Landlord and Tenant Act 1927, limiting claims on "repair" related damages claims has been superseded by common law. The Law Commission Report No 238 on *Landlord and Tenant: responsibility for state and condition of property* (1996), amongst other matters, looked at the issue of whether section 18(1) was even necessary anymore and decided:

> After careful consideration, we have decided not to recommend repeal at this stage, but to see how the law governing damages for reinstatement develops over the next few years. If it becomes apparent that section 18 is no more than declaratory of what the common law has now become (as we suspect it will), the section can then be repealed.

Since, in particular, the common law principles of loss, this Law Commission Report followed shortly after the leading judgment in *Ruxley Electronics and Construction Ltd* v *Forsyth* [1995] EGCS 117 in which Lord Lloyd of Berwick set out two key principles on loss that apply to dilapidations damages claims, regardless of whether or not the loss relates to a breach of repair obligations:

> first, the cost of reinstatement is not the appropriate measure of damages if the expenditure would be out of all proportion to the benefit to be obtained, and, secondly, the appropriate measure of damages in such a case is the difference in value, even though it would result in a nominal award.

Common law has also moved on further and in 2006 in case of *Latimer v Carney* [2006] 3 EGLR 13, the application of section 18(1) was further considered and the previous commonly adopted argument that decorations related items were not subject to any section 18(1) cap on "repair" related claims was examined. Arden LJ held that:

> Section 69 (of the L&TA '54) states that the word 'repairs' 'includes any work of ... decoration' ... section 70(1) provides that the 1954 Act and the 1927 Act may be cited together as the Landlord and Tenant Acts 1927 to 1954.
>
> ... In all the circumstances I consider that this court (the Court of Appeal) should treat a failure to repair the decorative state of the premises as a breach of the covenant to repair for the purposes of the first limb of section 18(1) of the 1927 Act even if that failure also constitutes a breach of a covenant for periodic decoration in the same lease.

It is clear that common law has moved on since 1996 to the extent that section 18(1) is virtually superfluous as common law now effectively provides the same degree of protection and "cap" on damages claims as section 18(1), but without the restriction of having to concern "repair" related claims.

Unreasonable range of professional opinion

While the courts view the measure of dilapidations breaches and damages arising from the breaches as a matter of fact, the only way of determining this factual position is to rely on the interpretation and expertise of the expert surveyors preparing and responding to the claim. Each individual surveyor's opinion will be a by-product of:

- the degree of their own personal relevant dilapidations experience
- the quality and primacy of the loss data and evidence available for consideration on each dispute.

It is therefore possible to have two surveyors read the same lease and view the same physical evidence of breaches and yet reach different opinions on the degree of severity of the breach and the appropriate remedial works.

In terms of the surveyor's experienced based opinion, any opinion formed should be reached on the basis of a careful and skilful interpretation and consideration of the lease obligations; and a comparison of those appraised obligations with the actual physical state of the subject property at the material date for the claim.

The surveyor's opinion and representations of their opinion should be that which a reasonable and competent practitioner could innocently determine but should not be so partisan in their client's interest that it becomes negligent or, worse still, is known to be untrue and so could be considered to be deceitful and possibly fraudulent.

In terms of the costs associated with the surveyor's opinions of the remedy, unless truly competitive tender figures are obtained from reasonably suitable and reputable contractors against an agreed scope and specification of remedial works, then any costs will also be subject to a reasonable "range" of possible costs.

Where *prima facie* cost evidence is not available and where price data books are used to value the works and help quantify the claim, care should be taken to be reasonable in considering the price data book published costs. In most cases, these published costs will be averages extracted from national statistical studies where the statistical "population" studies is not known and the standard deviation (s.d.) and quartile data is not made available.

As most cost studies follow the statistical "normal distribution curve", the reality is that the more appropriate cost value somewhere within the range of the mean average figure +/–2 s.d. (as this range accounts for 95% of the statistical population figures in normal distribution curves). In construction works, the s.d. can be hundreds or even thousands of pounds per measured unit and so the price book data may be practically worthless for skilfully measuring the true value and loss.

For each item in a dilapidations schedule, if the surveyor's reasonable range of professional opinion is combined with the reasonable range of cost data, the overall reasonable range of claim opinion can be found (see the illustration opposite p241).

During the negotiations stages, the more partisan the surveyor's opinions, the wider apart the difference of opinions will be and the more likely it is that the landlord and tenant parties will fail to settle the

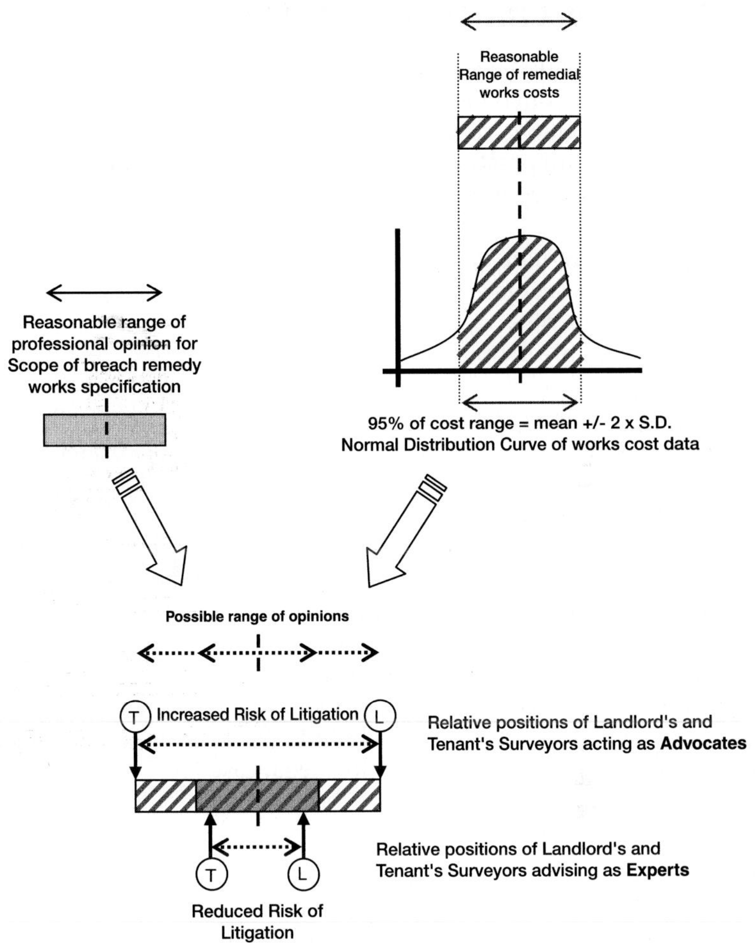

claim and proceed to litigation. Sadly, one party is likely to get a nasty shock as a consequence and the surveyor's professional indemnity insurance may get called upon.

However, if the surveyors recognise their expert role (see Chapter 4) then their opinions are more likely to be closer to the true and fair measure of loss and so litigation becomes less likely.

It should be every surveyor's objective to act in an expert and fair manner. The closer they are to the fair measure of any loss, the greater the litigation costs and court sanctions risk are to the other side.

Fixtures or chattels disputes

The legal demarcation between "fixtures" and "chattels" is an abstract area of law that is often highly confusing in the context of commercial property dilapidations disputes, particularly at lease end. The topic has many legal and practical applications and a clear understanding is required if the surveyor is to correctly interpret issues such as:

- reinstatement notices
- yield up covenants
- repair covenants
- conditional break options.

An appropriate interpretation of the issues will above all allow the surveyor working for either the landlord or tenant to accurately assess the extent of this area of the dispute. This will expedite settlement to the benefit of all concerned.

There is extensive case law relating to fixtures and chattels but this often lacks consistency and is based on an item-by-item review of the structure, fabric and contents of typical commercial buildings, covering issues that are not always applicable to every area of dispute.

An understanding of the basic legal principles applied in each case is required, following which each situation can be assessed. Where a particular element of a commercial building is the subject matter of the dispute, the surveyor should refer to a legal case law book to find the most closely applicable judgment.

Defining fixtures and chattels

What is the distinction between a fixture and a chattel? The surveyor must ignore the term "fitting" as this is a widely misapplied term and it has no relevance or meaning in law.

The term "fixtures" are defined as "part of the property" affixed to the demise, while "chattels" are defined as "moveable items" which never lose their separate identity. For example, a block-work wall is generally a fixture, while an office chair will be a chattel.

Whether an item within a commercial building is a fixture will depend firstly on the "degree of annexation", ie the extent to which it is fixed to the premises and the extent of permanence. Where an item is fixed to the property there is a rebuttable presumption that it is a fixture.

It is not always understood by surveyors that carpets are usually chattels and therefore may be removed by the tenant at lease end. Unless the landlord can absolutely prove, on the balance of probabilities, that asserts the carpets are owned by it, then on the basis that "he whom asserts must prove", unless the landlord can prove that the carpets are owned by the landlord, then the landlord's claim must fail. Carpets cannot be stated as owned by the landlord, unless provable; and that leads to a number of dilapidations claims, including carpet replacement, failing.

Secondly, the definition of a particular item as a fixture or chattel will depend on the reason for the installation. If it has been installed for the greater good of the property then it is likely to be a fixture. Conversely, if it has been installed in the property purely for reasons relating to the item itself, then it is more likely to be a chattel. Based on this definition a radiator on a wall would be a fixture, while a picture fixed above the radiator on the same wall would be a chattel. The first test should be considered indicative and the second test should be considered decisive.

These basic definitions form a sensible starting point for examining the areas of dispute that occur in relation to fixtures and fittings.

Differing landlord or tenant obligations

"Landlord fixtures" versus "tenant trade fixtures"— this is a legal subdivision of items classed as "fixtures". Landlord's fixtures cannot be removed by the tenant during the lease, but interestingly may have been installed by the landlord or the tenant. The basic principle is that if a tenant installs a fixture to the demise, then that fixture will become part of the demise and at lease end will be required to be handed back to the landlord, subject to the same covenants at lease end as the rest of the property.

In law there is the important exception of "tenant trade fixtures" that the tenant will be allowed to remove at the end of the lease and that the landlord will not be able to require to be left in covenant compliance. An example of this would be a commercial kitchen installation in a restaurant, factory or plant and machinery fixed to the structure of the premises in a mechanic's workshop.

The tenant is allowed this right of removal subject to making good any damage caused by the removal of the fixture.

Fixtures, chattels and vacant possession

The issue is not just one of academic importance. There are numerous practical implications, with perhaps one of the most important being a lease break clause in a commercial lease that is conditional on the tenant giving "vacant possession at lease end". Such clauses are surprisingly common and rely entirely on the distinction between chattels and tenants to define whether vacant possession has been achieved.

An example of such a clause might be: "Tenant to yield up the demised premises with 'vacant possession free from encumbrances' at the break date." The vacant possession requirement in the lease means that the tenant must return the premises to the landlord on the break date, with all tenant chattels removed from the premises. For example, even a chair left in the demised premises could give the landlord an opportunity to challenge the validity of the break.

The distinction is important, because as explained above, a chattel could prevent vacant possession, while if the object has become a fixture and therefore part of the demised premises, then vacant possession would not be prevented. Fittings might typically include demountable partition walls, false ceilings, and light fittings, etc if installed by the tenant. If a tenant fails to remove these items they may risk failing to deliver up the premises with the required vacant possession.

There is also an implied obligation in commercial leases to deliver up vacant possession at lease end.

Tenant fixtures at lease end

Where a landlord is happy to allow a tenant's fixtures to remain within a property at lease end, they may be unable to insist on this. As a general rule, the tenant will be entitled remove these fixtures installed at their expense at lease end. If the tenant opts to remove their fixtures, they must make good damage to structure, fabric, finishes and other landlord services, plant and equipment, etc disturbed in the process.

Alternatively, there may be terms in a lease or subsequent licence to alter and effectively permit the tenant to leave their fixtures, alterations and/or additions in the premises at lease end, unless the landlord specifically requests their removal (often to be requested subject to conditions requiring reasonable notice periods, etc). If the landlord then fails to serve a proper notice on the tenant before the lease end and is then left with the tenant's fixtures, the landlord may be unable to claim for the cost of any subsequent removal of the tenant's fixtures.

Another fixtures and chattels issue sometimes encountered is where a tenant has installed fixtures and chattels under an original lease; and at lease end then remained in the property under a second (or further) lease after the first lease expired.

Occasionally, the tenant's surveyors may claim that the tenant's fixtures were landlord's fixtures under the later leases, simply because they were present at the commencement of the second (or later) lease term. However, case law does not support this assumption with regards to tenant's chattels, which will remain the tenant's property until their chain of leases expires.

The ability to insist on removal and or reinstatement of the tenant's fixtures at the end of the chain of leases will be dictated by how the alterations were to be treated at the end of the first lease. If the parties have not agreed extensions of reinstatement obligation under the subsequent leases then the landlord may well have lost their right to seek reinstatement.

Where surveyors are unable to agree on fixture and chattel issues, they should consider taking a single joint appointed counsel's opinion where the value of the disputed elements of the claim justify the expense.

Schedule of condition disputes

The purpose of the schedule

Often when parties are negotiating a lease, one of the parties will look to protect their future dilapidations position with regards to the standard of, say, repairs or decorations, etc. This is normally achieved by arranging for a schedule of condition to be independently prepared for attachment to and cross-reference within the lease.

It should be remembered that the purpose of the schedule of condition is to record in absolute factual terms the nature and condition of the various structure, fabric and finishes elements making up the property. The emphasis is on a factual record and so the schedule should not include speculation or assumption.

Also, if the schedule fails to record or observe the condition of any part of the property, then the element lacking a condition record/comment will not benefit from any protection; and the relevant lease clause that the parties intended to be influenced by the schedule will apply unaltered for the particular item where no record of condition exists.

Furthermore, the need for independence during the schedule of condition preparation is essential, as both lease parties will rely upon it equally. It is not for the schedule of condition to seek to strengthen or weaken one or the other party's future dilapidations position, as the intentions of how the schedule will have a bearing on future dilapidations is for the parties to jointly agree within the drafting of the cross-referencing lease clause(s).

The intention of the schedule

When dealing with a schedule of condition, a common but sometimes erroneous assumption to make is that a schedule is only ever attached to the lease to set the upper standard of repair or decoration, eg that it is only ever attached to the lease in the tenant's favour, thereby removing the tenant's liability to ever put, keep or yield up the property in any greater condition. This type of intention is normally achieved by having the relevant repair or decoration covenant(s) obligation limited so that the tenant is required to put and/or keep the property in "no better" state than the standard as evident in the attached and referenced schedule of condition.

While the suggestion that the schedule may be intended to set an upper ("no better") condition standard may be true in the majority of cases; there are occasions where it may actually have been attached to the lease in order to set the minimum ("no worse") standard of repair or decoration, etc.

For example, a developer purchasing a listed property on sale and lease back terms where they grant, say, a five-year lease to the existing occupier, may want to be certain that they will get the property back in at least the schedule of condition state at the end of the term as they have already factored in the costs and any likely end-of-term dilapidations position during their purchase of the property.

This type of lower ("no worse") condition standard intention is normally achieved by having the relevant repair or decoration covenant(s) obligation defined so that the tenant is required to put and/or keep the property in "no worse" state than the standard, as evident in the attached and referenced schedule of condition.

The difference between the "no better" and "no worse" condition standards within an otherwise ordinary full insuring and repairing lease can be graphically illustrated as shown opposite on p247.

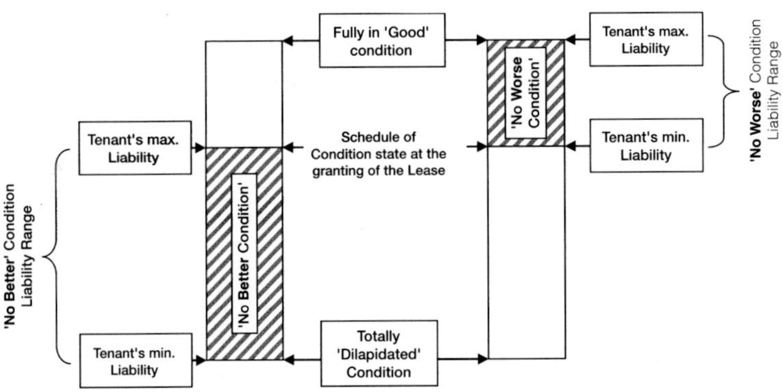

There may of course be occasions where the drafting of the lease failed to reflect the true or full intentions of both parties to the lease, or the relevance of the schedule of condition. However, in those circumstances, it becomes a matter for one party or the other to demonstrate alternative intentions and seek rectification of the lease before they will be able to widen the influence of the schedule of condition beyond the express influence contained in the lease as granted.

The relevance of the schedule

Another common misunderstanding is the frequent assumption that simply because a schedule of condition has been prepared, agreed and attached to the lease, that all the tenant's covenants and obligations will be influenced by the attached schedule. The truth is that the schedule of condition will have no relevance or material bearing on any clause (or other part of the lease), unless a clause expressly refers to the schedule of condition; and states how the schedule will affect the clause interpretation.

A further point to consider when interpreting the influence of a schedule of condition on the dilapidations claim is to consider whether or not any upper (no better) condition standard can be practically achieved. It is not uncommon to find that a tenant has taken sensible measures to protect their dilapidations position, only to find that the property condition has further deteriorated during their lease term and that the only practical way of addressing the deterioration is to undertake remedial works that exceed the intended lease covenant obligation.

Similarly, it may be that legislation or competent authority standards and requirements have changed or developed since the lease was granted and that a tenant seeking to undertake remedial works to a pre-defined standard is now unable to do so and is obliged by external legal influences to undertake works to a greater standard.

Careful interpretation

It could be said that sometimes surveyors fail to adequately appraise the relevance and influence of a schedule of condition attached to a lease when attending to a dilapidations claim. It is not uncommon to find that a surveyor simply notes that a schedule exists and immediately assumes it is in the tenant's favour and that it limits the possible claim arising under all lease clauses. However, in most cases, disputes between surveyors can be readily addressed if they impartially review the intentions of the schedule of condition and the relevance to the lease and permissible claim.

Loss of rent claims

How do surveyors currently assess loss of rent claims?

Scottish Mutual v *British Telecom* (1994) (unreported) is quoted in most text books and articles as the leading case for loss of rent claims. The case related to an office block in Ipswich, with a lease end in 1992 and at a time which saw very low levels of demand in the market cycle.

The experts for the landlord and tenant suggested a marketing period of two years and four years respectively. Importantly, both parties accepted that the contract period for the remedial works was just 20 weeks, considerably less than it would have taken to let the property.

Following the lease end, the case was litigated, with judgment given on a number of areas of dispute, including loss of rent.

The case established that where the tenant was in breach of covenant, the principal factor required for a loss of rent claim to succeed was that the remedial works procured by the landlord actually prevented the re-letting of the premises during the contract period. The judge in *Scottish Mutual* stated "Where loss of rent during the period required to carry out the repairs is to figure as a head of damage it is an essential prerequisite that it should be demonstrated on a balance of probabilities that the carrying out of those repairs after the end of the term has prevented or will prevent the letting of the premises for that period."

Landlords have subsequently struggled to recover damages for loss of rent in situations where a tenant is not actually lined up to take a new lease during the contract period. This is generally viewed as a helpful judgment for tenants and may be why such a high proportion of the loss of rent claims served on tenants fail.

The practical issue is evidencing the loss. Surveyors need to consider the following in relation to each claim.

1. Do you intend to carry out all of the works?
2. How strong is the letting market? Is it more likely than not that a tenant could be found now? Can you evidence this?
3. Will the dilapidated condition of the property affect/prevent marketing? Have you delayed the works or marketing unreasonably?
4. Is the loss of rent period due to non-dilapidation factors such as other works?
5. Do not mix a loss of rent for works period with rent free incentives.

The evidence

Landlords who have to carry out works in default of the tenant after lease end are often advised by their letting agents that marketing of the space is impossible until the works are complete, in all but the strongest market. In the case of office space, this would mean that the space had to be reinstated, repaired, redecorated and re-carpeted, ie until practical completion is achieved. Prospective tenants have repeatedly shown that they do not have the vision to see the potential in dilapidated premises, and that consequently the best rents can only be achieved when the property is in a state of good repair. However, is this enough?

If the landlord can show by arms length advice from his letting agents that local market conditions or the nature of the building are such that the property is effectively un-lettable until the works are completed, then a claim could be justified provided the landlord mitigated their loss by acting quickly and not including other works beyond remedies of the breach.

Strategies for surveyors acting for the landlord

- Landlord's works must relate to the breach. Additional works could complicate the claim.
- Bear in mind that the burden is on the landlord to prove their loss. The market must support the claim.
- The landlord must mitigate their loss and start works as soon as possible after lease end or even start preparations before then.
- The property should be marketed as soon as possible. Offering a rent-free inducement to a new tenant specifically to remedy specified items of disrepair may work but this must be evidenced to be repair specific.
- Obtain good contemporaneous agency evidence regarding the state of the market at the date of the lease end.
- If instructed at the start of the lease, consider whether an unambiguous loss of rent recovery clause could be inserted in the lease.

Strategies for surveyors when acting for the tenant

- Complete the works prior to lease end.
- Offer the landlord vacant possession prior to lease end for the period of time that the works will take to carry out.
- Give the landlord notice that there will be a breach, giving the landlord time to prepare the specifications and tender the works.
- Obtain contemporaneous agency evidence regarding the state of the market at the date of the lease end.
- Scrutinise the landlord works to see if there are additional works over and above those to remedy the breach.
- Carefully document the full extent of disrepair with photographs.

Value Added Tax (VAT)

Basic principles

The recovery of damages in lieu of VAT is a widely misapplied element of many landlord dilapidation claims, yet the logic behind whether damages are recoverable is relatively straightforward to establish.

The landlord's schedule should be a fair and accurate statement of the landlord's loss so that the tenant can understand the extent of the full liability if the tenant fails to complete the required works as soon as is reasonably possible. This is a requirement of the Property Litigation Association's (PLA) Dilapidations Protocol and essential information to settle any dispute.

When damages in respect of VAT are added to a claim, the inflationary impact of the dilapidations (in percentage terms) will be roughly equal to the current rate of VAT (17.5% at the time of publication). The misapplication of this element of the claim can create an embarrassing disparity between claim and settlement and may even delay settlement and the receipt of damages by the landlord.

HM Revenue and Customs (HMRC) take the view that lease end dilapidations payments are compensation for breach of contract, rather than a payment for the supply of goods or service, and are therefore outside the scope of VAT.

HMRC Notice 742 Land and Property, section 10.10 states the position as follows:

> A dilapidation payment represents a claim for damages by the landlord against the tenant's 'want of repair'. The payment involved is not the consideration for a supply for VAT purposes and is outside the scope of VAT. VAT may however apply to individual cost elements that form the basis of the claim for example survey and legal fees and the building works relating to the works that remedy the Tenant breaches. If the Landlord can recover the VAT element from HMRC then it cannot also make a claim for this amount from the Tenant at the end of the lease. Such a 'double claim' would go against the common law principle that the Landlord can only recover what he has lost. The courts would immediately reject this claim if litigated. Conversely if the Landlord cannot claim back the VAT from HMRC then he will be entitled to add damages in lieu of VAT to the claim.

When can the landlord recover VAT?

Having established the basic principles behind the recovery of VAT, the landlord's ability to recover VAT must be established in each case. The two-stage test is applied as follows:

1. Is the landlord registered for VAT at HMRC? Is the landlord required to charge VAT on normal business transactions such as rent?
2. If the answer is "yes" to the above question, has the landlord

waived the individual exemption for VAT on the specific building to which the dispute relates, ie is VAT charged on the rent and other "services"?

If the landlord has waived the VAT exemption on the building, eg VAT is paid on rent, then the VAT is fully recoverable by the landlord and damages in lieu of VAT should be excluded from the claim because the landlord can offset input against output tax and will not suffer this aspect of loss.

Summary

If a landlord charges VAT on the building, the landlord will be able to recover their own VAT when doing the relevant works. In this case, the departing tenant does not have to pay damages in lieu of VAT.

If the landlord has not elected to charge VAT, then the tenant will have to pay damages in lieu of VAT on the dilapidations claim if the landlord is actually going to do the works. If the landlord is not doing the works, eg they are selling or just passing the dilapidations payment to an incoming tenant as a rent-free period, then the damages received would be classed as a capital receipt and tax would be paid to the HMRC.

The tenant cannot force a landlord to charge VAT on their building just to save the tenant paying it. The option to waive the exemption for VAT on any building is entirely a landlord's prerogative.

Tenant's lease "break" disputes

Break notices included within the lease create a contractual entitlement for one party (or both) to terminate the lease prior to the lease end date stipulated.

The starting point with break notices is that surveyors should not serve a schedule of breaches unless they are certain that this approach accords with the type of break clause in question. The termination of a lease by a tenant can often be difficult for the tenant to operate, if the strict conditions are not complied and the tenant may potentially remain liable under the lease for the remainder of the term. In some circumstances this may be a significant benefit to the landlord and the landlord need not proactively assist the tenant in the operation of the break.

Conditions set out within a tenant's break option can include requirements to comply with one or more of the following:

- A specified time period of written notice which terminates on a specified date.
- Payment of rent and other sums payable under the lease up to a specified date, often the break date.
- A requirement for the tenant to yield up the demised premises with, eg "vacant possession free from encumbrances" or in compliance with certain lease covenants on a specified date. Where full compliance with the tenant's covenants is included as a condition precedent, then the right to break can be lost on extremely narrow and technical grounds.

Working for the tenant when works are required to operate the break

If the break clause requires that certain works should be undertaken, a prudent tenant attempting to operate the lease break should agree the scope of works with the landlord, be prepared to vacate the premises, and should have arranged for works to commence in time to comply with the break. The extent of works will depend on legal interpretation of the relevant clause.

It is extremely important to set a firm longstop date for the dilapidations negotiations. The purpose of the longstop date is to allow the tenant sufficient time to carry out the repair and redecoration works to the premises if negotiations with the landlord stall.

The tenant's surveyor should prepare a list of all works that are required to address all relevant breaches. The works must comprise all works required, and the list should be served on the landlord with a request that the scope of works is commented on and agreed if possible. It may even be prudent to offer to pay the landlord's surveyor's fees for this process.

The list of works prepared and issued by the tenant should be in a format that can easily be understood by the landlord's surveyor, ideally a Scott schedule format.

Working for the landlord when works are required to operate the break

Landlords can often reasonably adopt a "wait and see" policy pending advice and clarification from their client's lawyer and also following their review of the lease and the approach taken by the tenant.

There is scope for disagreement between the landlord and tenant over extent of the tenant's obligations on the break clause. Such disagreement could be very costly for the tenant if it results in court action. Lawyers should provide a second opinion and to maintain a "watching brief".

Mechanical and electrical building services and dilapidations

Introduction

Mechanical and electrical (M&E) services are a hugely overlooked element of dilapidations disputes that can equate to 50% of a claim. Many leading court cases have M&E services at their core. Because nearly every commercial property will have some element of electrical or mechanical installation, it is very likely that this element will be involved in the lease end claim and the resulting dispute.

The dilapidations surveyor must firstly learn to spot the potential areas of the claim that can involve M&E. Secondly, the surveyor must understand how to deal with them and when to recommend to the landlord or tenant client that they should appoint a mechanical or electrical specialist to address the issues.

There is a paradoxical relationship between dilapidations and M&E services, ie claims generally benefit from M&E input; and leases generally allow recovery of the engineer's fees; but generally building surveyors and lawyers do not use engineers when preparing or defending the claim. Surveyors and lawyers specialising in dilapidations should be alive to the opportunities that engineers can offer to bring about a quicker and fairer settlement in accordance with the Civil Procedure Rules (CPR) and the Royal Institution of Chartered Surveyors (RICS) Guidance Note.

Where do M&E services fit in?

Any particular M&E services element of a commercial building can be the subject of the dilapidations claim itself, or it can have an impact on another area of the claim. For example, if the air-conditioning system is demised to the tenant and it is in disrepair at lease end, then it will directly form part of the landlord's claim against the tenant who was responsible for maintaining the installation during the lease term. Alternatively, if the air-conditioning system is not demised and is out of repair at lease end then works that the landlord has to undertake may impact significantly on the tenant liability.

By simply considering the two scenarios above it is essential to establish:

- who is liable for repairs
- what elements of the building (including services) fall within the tenant's repairing obligation
- what is the required standard of repair.

Commonly encountered M&E installations/issues in dilapidations claims include the reinstatement and/or repair of:

- electrical wiring
- lighting
- distribution boards
- power points
- cold water supplies, pipes and tanks
- heating apparatus
- air-conditioning plant
- fire alarms
- specialist manufacturing plant
- drains
- statutory compliance
- lifts, etc.

The engineer's role for each item will focus on either the reinstatement of a tenant alteration, or the remedial works to address the disrepair.

The landlord or tenant surveyor needs to be able to explain (to themselves and his client) whether one or two days of an engineer's time is justified on any given job. This is a judgment based on the complexity and value of the claim.

There can be several reasons for surveyors to have M&E services dealt with by an independent specialist. These include:

- M&E installations often have a shorter life than the building fabric they are set within. Therefore, M&E problems tend to arise before others.
- They are often out of sight.
- Once disrepair is established the position can be complicated because an exact replacement may be impossible to source, or replacement may give rise to a considerable improvement which can initiate further legal argument.
- An exact replacement of original plant may be against modern regulations or it may be so out of date that no reasonable landlord would want it.
- Most surveyors are not qualified to comment on M&E services in any detail. The right engineer will be.

Practical tips when dealing with M&E issues

Engineers can add:

- greater clarity and detail to the claim
- improve the surveyor's understanding of the premises and the claim
- give more accurate costings
- speed up the resolution of the dispute. The schedule is still reasonable, will remain CPR compliant and is more accurate.

Explain to the engineer the nature of repair or the reinstatement required

The engineer needs to be guided on the legal liability defined in the lease which will determine the engineer's approach in terms of stating the appropriate remedy. The surveyor or lawyer instructing must remember to keep the instructions clear and cover the basic principles if the engineer is not widely experienced in commercial dilapidations.

Reinstatement

The impact at the end of a lease of tenant alterations during the lease depends largely on the terms of the lease and any licences that the landlord granted regarding alterations. On granting consent for alterations, the landlord probably required that at the end of the lease the tenant must "restore the property to its original state if requested to do so". However, this is not always the case so engineers must check before proceeding.

Alterations nearly always involve an M&E element, and are often some of the more complex matters that surveyors deal with. The input of an M&E engineer can be crucial.

Repair

Many commercial leases will include a covenant "put and keep the property in repair" or similar. Unless the tenant and the landlord specifically agree otherwise, the fact that the premises were in a poor condition at the start of the lease is sometimes irrelevant. The tenant still has to put them right and this relates to M&E as much as anything else.

M&E often involves major capital expenditure, for example on lifts, air-conditioning, and heating. In each case, the question must be: "Can the particular item be patched up, or can 'repair' only be achieved by wholesale replacement?" To fall within a dilapidations claim the equipment must be in disrepair. It is not enough that it is out-moded, obsolete or old fashioned.

Plus, in each case the tenant is likely to point to the impact of the disrepair on the capital value of the property, ie section 18 of the Landlord and Tenant Act 1927, aka the "statutory cap" on damages. If the particular item of disrepair relates to M&E then this will increase the importance of the engineer.

Unless guided clearly, engineers often assume the worst and replace or repair substantial elements of the services installations. An engineer's experience may be based on Cat A refurbishments. If this approach is applied to the preparation of a landlord claim, it can easily come undone on close examination in the dilapidations process.

Surveyors must understand the legal and theoretical basis of the claim and communicate this prior to the engineer's physical inspection. The actual result of the site inspection must evolve through a process of dialogue between the engineer and surveyor.

Case law has established that for disrepair there must be:

physical damage or deterioration resulting in non-compliance with the standard contemplated by the covenant. An item of plant that is old fashioned or inefficient is not automatically in disrepair.

The landlord can only recover the cost of work from a tenant that relates specifically to a tenant's breach of lease covenant. A common question is whether the landlord can recover the cost of full replacement of plant, or just piecemeal repair? To answer this question requires careful judgment by the engineer and the surveyor, often with a lawyer's input.

The system has to be in good working order. It must be in a state of repair and work substantially as well as the original system did when new. However, the plant does not have to be as good as new, nor does it have to require as little maintenance as a new system.

It is often the responsibility of the engineer to advise their client on the option of repair or replacement. If the engineer can reasonably argue that either course of action is appropriate, then (if the courts agree) this is acceptable.

Another common question is whether the landlord can replace old plant with modern. Again, careful judgment by the engineer, surveyor and lawyer is vital. Because M&E installations have a shorter life expectancy than the building fabric, this question arises very often in dilapidations disputes.

There is substantial case law on the topic and input from a lawyer will often be required. Generally, the position is that replacement will constitute repair if the new installation is "substantially similar to the old installation".

Always remember that each case will turn on its particular facts. Engineers must be asked to exercise caution and question the conclusions they offer at all stages of a dispute. Never hesitate to consult the client and their lawyer. Always read the lease.

Understand what an M&E engineer can do

Whether acting for landlord or tenant, engineers can have varying degrees of input on a dilapidations dispute. When acting for the landlord or the tenant on a full lease end claim for dilapidations, the engineer's role will include:

- Surveying the extent of the mechanical and electrical services within the premises to establish what elements are a landlord or tenant's responsibility. The demarcation between landlord and tenant responsibility is a common mistake in schedules served by landlords.

- Reviewing the service and maintenance records.
- Reporting on the state of repair of the M&E installations and reporting on their condition reference to the lease with a clear, accurate, and impartial written report, using the appropriate dilapidations schedule format.
- Reviewing licenced and unlicenced alterations.
- Specification of remedial works to prove the loss.
- The surveyor must make an early evaluation on whether the engineer is required.
- Whether the cost is recoverable under the lease and if applicable, get the engineer involved early. If instructions go to the engineer they should set out clearly what is required.

Engineers must be independent from the start

Engineers need to be guided but they must remain independent. The terms of appointment need to be clear and open and stress the independent CPR compliant role that they must take. Whatever role is taken on by the engineer, they must consider their duty to be an independent "expert".

Engineers are involved in a claim for damages, ie for a breach of contract. With this comes the added responsibility of their "over-riding duty to the court". This duty is in place right from the start of the claim, ie before litigation. For details about expert witnesses in court, please refer to Part 35 of the CPR and speak to the lawyer involved in the claim.

Engineers have a duty of impartiality to flag up the strengths and weaknesses of the landlord claim or tenant response. Surveyors should use a clear letter of appointment to the engineer, one that they and their clients would be happy to disclose to the other side if requested.

The CPR came into effect in April 1999 and had a major impact on the way in which dilapidations disputes are conducted in the courts of England and Wales. Numerous areas have been impacted, many of which affect the work of engineers. An issue particularly relevant to the M&E engineer is the *Pre-Action Protocol* — a significant component of the CPR which established standard procedural guidelines for all persons involved in dilapidation disputes, including engineers. The Protocol sets out a detailed timeframe for negotiations from and including the service of the schedule. Engineers just need to be aware of the Protocol and seek guidance from the surveyor or lawyer.

Resolve the M&E paradox

There is a paradoxical relationship between M&E engineers and dilapidations claims. If surveyors understood this they would resolve dilapidations claims more efficiently.

The paradox is that too few surveyors and lawyers consult M&E engineers on the preparation of a schedule of dilapidations, despite the fact that:

1. The vast majority of schedules of dilapidation cover M&E issues (around 80%). M&E elements regularly make up 40% of construction contracts and dilapidations is no different.
2. Only a small minority of surveyors use M&E Engineers to prepare the schedule — only 10–20% of schedules have specialist M&E input.
3. The vast majority of leases will allow the landlord to recover the costs incurred regarding an M&E survey if it can be justified — which in most cases it can!
4. The majority of claims would benefit from M&E engineers' expertise.

The CIBSE schedule of life expectancies

The Chartered Institution of Building Services Engineers (CIBSE) guidelines are extremely useful, but engineers must be careful when applying them to dilapidations claims. Take account of them but do not be bound 100% by them. The physical/visual inspection must always take precedence.

While the demise that you are examining on behalf of the landlord may not be high spec' and may be capable of substantial improvement to bring it up to current standards, it is the physical condition of the installation reference to what the lease requires that is relevant. For example, a 15-year-old gas boiler (in imperfect condition, but serviceable and functioning) is due for replacement according to CIBSE guidelines. Arguing replacement based just on CIBSE guidelines would not justify a landlord dilapidations claim.

Summary and conclusion

If surveyors use an engineer on the lease end schedule of dilapidations (or tenant response), they are more likely to produce a schedule with detail on key issues that will help prove the claim and comply with statutory obligations. This would result in quicker and more accurate settlement due to greater clarity and detail to the claim, which in turn would result in a more efficient dilapidations dispute resolution process.

Experienced dilapidations engineers rate the existence of detailed service, testing and maintenance records as the single most important indicator of potential breach of covenant. If they do not exist, the alarm bells will start to ring. If they are present and well documented they will reveal far more information than the sometimes brief dilapidations site inspection.

16 An Introduction to Service Charges

An overview of service charges

What are service charge disputes?

Disputes about service charges arise mainly under two circumstances:

1. Disputes between two or more tenants in the same office building or shopping centre, concerning the contributions levels paid by the tenants towards the running costs of the building or shopping centre.
2. Disputes between one or more tenants and the landlord, concerning the landlord's request for contributions to certain items, especially the replacement costs of air-conditioning, roofs or lifts or the removal of items such as asbestos.

Why are there fewer service charge cases?

One of the main differences between a service charge dispute and a dilapidations dispute is that there are far fewer service charge cases that have come before the courts. As we have seen earlier in this book, case law on dilapidations disputes goes back a few hundred years, however, service charge case law goes back less than 50 years.

Compared to dilapidations, service charges are a recent addition to the differences of opinion that may exist between a landlord and a tenant.

When there were fewer multi-let premises and prior to the construction of shopping centres (whether enclosed or open to the elements), there would have been fewer landlords and tenants whom could have disputed the costs of providing common services.

However, dilapidations disputes have, as we have seen earlier in this book, existed for over 700 years. In any lease of premises where a landlord obliges a tenant to keep in repair, redecorate or reinstate those premises, whether internally only or on full repairing terms, there is always the possibility of the parties disagreeing on how those items should be carried out or what the costs should be if the landlord is carrying out those tasks after lease end.

On the other hand, service charge disputes cannot exist where premises are let to one occupier and there are no common parts; nor where the rent is inclusive of the costs of providing the services (an "inclusive rent").

One of the best known cases which came to court involving a service charge was that of *O'May* v *City of London Real Property Co Ltd* [1982] 1 EGLR 76. Put simply, the landlord wished to include a service charge obligation in the renewal of a lease, for the first time. The tenant, perhaps not surprisingly, objected to this on the basis that they had not had any service charge obligations prior to that in their existing lease. This case, one of several quoted in the Royal Institution of Chartered Surveyors' (RICS) Service Charge Code, in force from April 2007, led to the principle that the terms of a lease being renewed should follow the terms of the existing lease. It is for the party wishing to change the terms of that existing lease to prove, if it can, that the changes are fair and reasonable.

Dilapidations disputes, of course, occur in both full repairing leases of single-let buildings as well as internal repairing leases of multi-let buildings, with the external repairs paid for via a service charge.

Another reason may be that a dilapidations dispute commences with a landlord serving a schedule of dilapidations on its tenant, whereas a service charge dispute is with a tenant challenging some or all of the items that it is being requested to pay. When one looks at the role of the landlord, it is clear that the landlord is in the business of dealing in property and therefore is likely to wish to maintain or improve the value of their investment. One of the ways to do this is to ensure that former tenants carry out (or pay towards) the repair, redecoration and reinstatement works required to that investment. It is highly likely, therefore, that a landlord would serve a schedule of dilapidations on their building, unless the tenant has complied with each of the lease's repair, redecoration and reinstatement obligations.

Most tenants are not in the business of property and so would not particularly notice that some of expenditure requested by the tenant are, or would be, in excess of what should be paid. As the tenants would not know how to challenge items that may appear not to be in order, it is only those items that are particularly excessive that would make it worthwhile for a tenant to bring in solicitors and surveyors to advise on the position.

This differential between the two types of disputes inevitably leads to far fewer cases coming to court, to create the case law as guidance for practitioners in this field compared to dilapidations case law.

Finally, there may be another reason for fewer service charge cases compared to dilapidations cases that have gone to court; the fact that a landlord could threaten or use, without notice, bailiffs to collect disputed service charge sums which had been withheld. The fact that a landlord can currently use bailiffs to recover disputed service charge sums has lead to tenants "caving in" when they are made aware that a landlord is considering that route available. It is thought that tenants will be in a stronger position when that route is closed off by this statute.

The change that is about to happen, that will stop landlords using bailiffs, is the Tribunals, Courts and Enforcement Act 2007. Although it was enacted in 2007, this statute has not yet been brought into force. However, when it is in force, it will remove the landlord's right to consider bailiff action over disputed service charges which have been withheld. The Act is likely to be brought into force in early 2009, when the Regulations that support the statute will have been agreed and published.

Damages versus debt claims

Whereas a claim for dilapidations by a landlord against its tenant or former tenant is a damages claim, a service charge dispute is merely a debt claim.

Due to the fact that a landlord does not have to prove a loss (as a landlord does with a debt claim), there are fewer dilapidations cases. Therefore, there will always be fewer service charge cases coming to court than dilapidations.

Additionally, many of the dilapidations cases that come to court involve matters that are not part of a service charge claim, such as arguments about the correct diminution in the value of the landlord's interest, arguments about supercession meaning that various items should not be paid for by the tenant/former tenant and no loss, at all, should be proven where the building may be redeveloped or demolished entirely soon after the lease end.

Conversely, the only type of dispute that a service charge case could include would be whether items should be repaired or replaced, whether contributions towards items sought was or was not the tenant's liability if the tenant enjoys an exclusion from those costs in the lease, whether the apportionment of the costs have been correctly carried out and whether works carried out by the landlord were done during the time that a tenant had a liability to pay towards them.

In summary, it can be seen that there is a far higher potential to have a dispute over a dilapidations claim than over a service charge claim, even where similar costs may be involved.

Statutory provisions governing service disputes

Unlike dilapidations disputes which have a number of statutory provisions governing what can and what can't be recovered from a tenant, listed earlier in this book, there are no statutes at all which govern service charge disputes involving commercial premises.

This book details the statutes that regulate residential property, but where premises are wholly commercial, with no part residential, the only assistance which practitioners have to guide them on what should be paid and not paid is the small number of cases which have been heard at the High Court.

Therefore, there is a "free market" in commercial property service charges, unrestricted by any statutory provisions. So long as a landlord seeks to recover what they have truly spent, or will spend, on a property, then the tenant is obliged to pay these sums.

There are a small number of exceptions that restrict full recovery from a tenant or tenants. These will be looked at more closely in Chapter 18. The exceptions are as follows:

1. Where the landlord has agreed to a cap on the service charge in the lease documentation.
2. Where the landlord has agreed to an exclusion in the lease documentation (such as no sums need be paid towards planned lift replacement or asbestos removal).
3. Where the lease is about to expire and the landlord is planning large capital works (case law restricts such full recovery).
4. Where there is an argument about whether items can be merely repaired, at a lower cost, and, debatably, do not need replacing.

17 Commercial Service Charges

Commercial property and the RICS Service Charge Code

Is compliance with the Code compulsory?

The Royal Institution of Chartered Surveyors (RICS) Code of Practice: *Service Charges in Commercial Property* is not compulsory and departures from the Code are permitted where reasonable.

However, as the Code is a RICS guidance note, it should be adhered to by RICS members, unless one or both of the following occurs: either the client determines that the Code will not apply in the particular circumstances relating to the property concerned; or the surveyor has determined that there is good reason for that departure.

Where departures from the Code are permitted

As the Code is not based on any statutory requirements (only residential tenants enjoy statutory protection), it cannot be forced on the landlord or the tenants.

The landlord or the tenants of a particular building may decide that the recommendations set out in the Code should not apply to that building. They are free to do this if they consider that the management of the building can be achieved in a different way from those recommendations set out in the Code.

Therefore, a landlord or a tenant may instruct their surveyor not to abide by some or in fact any, of the recommendations in the Code. If that occurs, then it would be good practice for the surveyor to have that instruction confirmed in writing by the client, as this would assist the surveyor in any challenges that might be made by the other parties' advisors, especially where the surveyor is acting for the landlord.

Additionally, the surveyor may consider that a departure from the recommendations may be preferable in relation to a particular building. In that case, it is advisable that the surveyor makes detailed notes as to why they consider that such a departure is preferred.

Sanctions for non-compliance

There may be sanctions against a surveyor who does not abide by the Code and that is why it is advisable for the surveyor to obtain written instructions from the client confirming that a departure should be made from the Code or that the surveyor has made detailed notes as to the reasons for departing from it, if it is the surveyor's view that such a departure be made.

If a surveyor has a professional negligence claim made against them, it is likely that a court would look at whether or not there are guidance notes that cover the point at issue. If the negligence claim concerns service charge payments, then clearly this Code will be considered by a court as being relevant to the points at issue.

The fact that the surveyor has written confirmation from their client not to abide by some or any of the recommendations in the Code, will assist the surveyor in defending a claim of professional negligence. Similarly, the fact that the surveyor has made detailed notes concerning the reason why the surveyor has determined to depart from the Code's recommendations, will show that the surveyor has given consideration to the recommendations in the Code, but decided to pursue an alternative way of managing the premises. Without those notes, it would be harder to defend a claim of negligence as it may show that the surveyor has not read the Code or had read it and ignored its recommendations; neither of which would show the surveyor in a positive light.

Guidance note for RICS members

The fact that the RICS determined that this Code should have the status of an approved guidance note gives the Code higher status than the two previous guides. It is officially endorsed by the RICS and, with it being a guidance note, it means that the RICS will expect its members to abide by it, unless there are reasons (set out above) as to any departure from it. In the event that an RICS member is investigated by the professional standards committee, the fact that a member has, or has not, abided by this guidance note may be relevant.

There are those permitted reasons for not abiding by the Code, but unless a detailed consideration has been recorded on the surveyor's file or written confirmation has been received from the client concerning departing from the Code's recommendations, then the surveyor may not be in the best position when defending any allegations made where the RICS have been asked to investigate matters.

Position of Code with replacement of plant and equipment

The Code is designed to minimise the possibility of a dispute between landlords and tenants. It states that it is to embody "best practice" and the RICS consider that surveyors and practitioners who abide by the Code should follow best practice in dealing fairly with running the common parts of commercial buildings and collecting sums from the tenants to pay towards these costs.

It is interesting, therefore, to note that the Code follows a different path in relation to replacement of plant and equipment than the current case law precedents at the High Court and Court of Appeal.

In paragraph D2 of the Code, the recommendation is that where plant and equipment serving the building is nearing the end of its useful life and requires replacement, the item may be recovered through the service charge. The example the Code gives is that of a heating system that is at the end of its working life which cannot be replaced "like for like" due to technological changes. There cannot be an identical replacement of the system due to it not being manufactured anymore. The Code states that it is common sense to include the costs of minor improvements to the plant and equipment of a property even where the lease only permits items to be repaired. The Code goes on to quote part of the judgment of a case from 1956, *Morcom* v *Campbell-Johnson* [1956] 1 QB 106, where Lord Denning had said:

> If the work which is done is the provision of something new for the benefit of the occupier, that is, properly speaking, an improvement; but if it is only the replacement of something already there, which has become

> dilapidated or worn out, then albeit that it is a replacement by its modern equivalent, it comes within the category of repairs and not improvements.

Finally, paragraph D2 recommends that where works go beyond the minimum specification required to repair an item, and this additional cost can be justified in terms of reduced maintenance costs, then that cost can be recovered through the service charge.

This view is in contrast with the recent case law at the High Court and the Court of Appeal. There have been half a dozen cases at these courts where it has been generally held that tenants should not be paying for anything other than repair of an item or replacement of the same.

Fluor Daniel Properties Ltd v *Shortlands Investments Ltd* [2001] EGCS 8 concerned the replacement of an air-conditioning system in an office building. The system was not in disrepair but the landlord wished to upgrade the system and claimed £2 million from the tenant. The tenant objected to paying this sum as it argued that the system was not in disrepair.

The High Court held that the full costs of these works could not be recovered from the tenant under the service charge. The wording of the lease presupposed that there would be some disrepair to the system or a defect in it before works could be carried out to replace it. The judge held that there was not any disrepair or defect and the landlord would not be able to recover these costs.

Part of the judgment is worth quoting, as it states that works proposed by the landlord must:

> presuppose that the item in question suffers from some defect (ie some physical damage or deterioration or, in the case of plant, some malfunctioning) such that repair, amendment or renewal is reasonably necessary ... Whether, once those conditions are established, the item must be repaired or renewed is a question of fact and degree having regard to the nature and extent of the defect and, not least, to the costs likely to be involved.

It is interesting to compare and contrast this case with the recommendations in the Code. This case said that an item could not be replaced unless and until there was a defect in it or it was in disrepair; whereas the Code recommends, in paragraph D2, that an item may be replaced in the circumstances set out above. This may be due to the Code being designed to assist landlords and tenants in the running of commercial buildings and avoiding possibilities of conflicts between the parties. However, where a tenant holds the view that they should only pay for the maintenance and repair of items (and nothing more) then, depending on the precise wording of the lease, it is possible that there would be a dispute between the parties as to liability for these costs.

Additionally, the High Court judge in *Johnsey Estates (1990) Ltd* v *Secretary of State for the Environment, Transport and the Regions* [2001] 2 EGLR 128 held that if an item is technically obsolete but not in the position of being in disrepair, it need not be replaced. This was a dilapidations case rather than a service charge case, but the same principle applies in relation to whether an item needs to be replaced at the tenant's expense.

In the case of *Scottish Mutual Assurance plc* v *Jardine Public Relations Ltd* [1999] EGCS 43, the tenant occupied a floor in a multi-let office building. The landlord was responsible for repairs to the common parts which included the roof. During the term of the tenant's lease, the landlord carried out patch repairs but, close to the end of the tenant's lease, the landlord carried out substantial works to the roof and sought recovery from the tenant.

The case came to court and the landlord was not able to recover the full amount due, on the basis that the tenant's lease was close to its expiry and the judge considered that the tenant would not enjoy the benefit of those works. This decision was based on the tenant having a reasonably short lease (three years).

Care should be taken by the landlord to ensure that service charge works are carried out while the tenant(s) is/are still liable to pay towards them. This appears an obvious point, but it often happens that tenants' leases end while works have not yet started and often the landlord will have to bear those costs themselves. It is not unknown for managing agents to face professional negligence claims from landlords where the works are planned but have not started and the tenants have left the building and, with their leases having expired, are "off the hook" in having to pay for them.

In the case of *Capital & Counties Freehold Equity Trust Ltd* v *BL plc* [1987] 2 EGLR 49 that is exactly what happened: the landlord had signed the contract with the contractors to carry out works to the common parts of the premises; however, no works had started at the time of the tenant's lease end. The court held that the landlord could not recover these costs as the tenant was only liable to pay for services provided by the landlord during the term of the lease and not after lease end.

Another problem a managing agent or landlord faces is where a lease originally granted on full repairing terms to a tenant is renewed but the renewed lease is only on internal repairing terms. This may happen where the negotiating strengths of the parties leads to the tenant not having to carry out works to the mechanical and electrical (M&E) equipment in the building, but to pay towards it via a service charge. Again, it does happen that the agents or the landlord miss the point that works to the "big ticket" items like the lifts and the air-conditioning can only be recovered if they are carried out during the lease term.

The landlord may instruct their surveyor to prepare a schedule of dilapidations for service on the tenant on the basis of a full repairing term of the lease; when they should have carried out the works to the M&E during the lease term and dealt with the internal repairing, redecorating and reinstatement items either before or after lease end; but not be under the same time pressures that they are under with the service charge items.

The surveyor in this case will have to carefully consider the lease terms on instruction and advise the landlord promptly in order to avoid this costs shortfall if works to the M&E are not carried out during the term.

Appointment of the service charge surveyor

The procedure for appointing a surveyor to act on a service charge dispute, the considerations as to the type of fee basis and the role of that surveyor (whether expert or advisor/advocate), are identical to that of a dilapidations appointment.

However, there is a slight difference: whereas a surveyor about to be appointed on a dilapidations case has to abide by the RICS guidance note on that appointment, there is no such guidance note or practice statement that governs that appointment when acting on a service charge dispute. The RICS Service Charge Code does not extend to governing the appointment of a surveyor to act on such a dispute.

It is taken as read, though, that a surveyor about to be appointed on either type of dispute will be subject to the applicable practice statements and guidance notes of the RICS; namely: the Practice *Statement, Surveyors Acting as Expert Witnesses, or the Guidance Note, Chartered Surveyors Acting as Advocates*. These publications govern those roles and the fee basis that may be agreed, depending on whether a surveyor is acting as an expert or as an advisor/advocate. However, the fifth edition to the RICS Dilapidations Guidance Note also covers the appointment of the surveyor and how they should act for that client.

18 Procuring and Apportioning Service Charges

Audit trails

Tenants are often able to challenge landlord requests for payments on large one-off works to plant and equipment by arguing that the landlord has not gone about the arrangements for procuring the works correctly.

For example, when a landlord is planning to replace the lifts in a building, the landlord may instruct the managing agents to ask the usual lift contractor who maintains the lift to quote for these works.

The contractor, seeing that a small annual maintenance contract could turn into a large replacement instruction, will put forward their quotation for this work and the managing agent would make their recommendation to the landlord.

All this seems acceptable to the landlord; after all, the lift contractor who knows the lifts well from their experience of maintaining them over the years, is being asked to quote for the replacement of the same. A landlord may think that no tenant would wish to object to this.

However, tenants quite rightly do object to this type of procurement. It is not seen as dealing with the repair or renewal issue independently.

The better way of procuring this type of work is for the landlord to create a proper "audit trail" that may be shown to the tenants. Tenants are less likely to object to major works being carried out (and, of course, to be paid by them) if it can be shown to them that the landlord went about arranging for the works in an independent manner. This includes not being swayed by the fact that it is in the interests of the building owner that large items of equipment should be replaced rather than merely repaired, if tenants are paying the costs of such work!

Independent reports

A landlord may consider that there could be nothing for a tenant to object to if they arrange for the maintenance contractor (who knows the item of plant or equipment better than any other firm) to quote for replacement works. What could be wrong with that, the landlord may think.

The problem for the landlord is that the tenants may be able to prove that not only did the work not need to be done (eg the lifts or air-conditioning system could just have been repaired), but the works may have been obtained from a different contractor at a lower price than the maintenance contractor charges for the same work.

Therefore, the best way of overcoming this problem would be for the landlord to obtain an independent report from a consultant who is an expert in the field of the item concerned, be it lifts, air-conditioning, roofs and so on.

This consultant should be wholly independent of the instructions that the landlord would give if the works were to go ahead. Often, landlords start with good intentions of commissioning reports from consultants on replacing items of equipment and plant, in order to prove to the tenants that the works are needed. However, they then fail in what they are trying to achieve when they let the consultants know that if they recommend replacement of the items in question, that firm will be promised the project management of the work!

It is only natural business sense that a firm which are specialists in the business of supplying and installing equipment and plant to offices and shops would be swayed by the fact that there is more money to be earned by recommending replacement, rather than repair of the item they are reporting on. It is not a bad reflection on air-conditioning, lift and escalator companies, etc that they may be influenced by the thought that additional fees, far higher than the standard report fees, could come their way if they considered that it would be better to replace an item rather than keep maintaining and repairing it.

Competitive quotes

Carrying on with the procedure of creating an acceptable audit trail for the tenants to have copies of, the independent consultant should be asked by the landlord when the report has been considered, to obtain three competitive quotations based on a precise specification for the works concerned.

In order to maintain the independence of the consultant, the three parties which would be asked to quote for the work should not include the consultants themselves. If they do, it could cause potential conflict of interest challenges from the tenants.

The landlord should ask the consultants to recommend which of them should be instructed on the work that the consultants have advised on. The landlord should bear in mind that it would not necessarily be the cheapest of the three quotations that is chosen. However, the tenants may consider that if the consultants had chosen three firms in order to seek quotations, then it would be unusual if the consultants would be recommending anyone other than the cheapest, if those firms are on an equal footing in the eyes of the consultants.

Full disclosure to tenants

It would be preferable to let the tenants see all the quotations and the consultants' report when letting the tenants know what works are being planned. Tenants may ask for this detail, in any case, but by passing these items over to the tenants as a full package, will make the tenants feel that they are involved with the decision-making process. The landlord is likely to get the full agreement of the tenants if they see all papers on an open book basis.

If the tenants do not see the report, quotations and reasoning for the works, they may feel that the landlord is merely planning to carry out replacement works at their expense.

Having tenants "buy into" the works

If landlords wish to be as sure as they can be that tenants will agree to pay for large, one-off works, including, but not only, the "big ticket" items referred to above, they can do this by having the tenants "buy into" the idea of the works.

Good communication

Regular communication between landlord and tenants about all aspects to do with the running of a building will mitigate the possibility of a dispute between the parties on what works should be done and/or who should pay for them.

Tenants will be hard pressed to argue that works should not be paid for if they have been properly consulted about them from the initial stages when the works were being planned, right through to the quotation stage and then the carrying out of the same.

Regular landlord-tenant meetings where large items of expenditure are planned and discussed will improve relations between the parties and let the tenants feel that they are being consulted on the running of the premises and the expenditure that they will be asked to pay towards.

This is all the more important when landlords are seeking to replace items such as escalators in a shopping centre or, say, air-conditioning in an office building when they are not in disrepair. These items would, very often, not be capable of being replaced at the tenants' cost as they may still be repaired and maintained, although they may be starting to become unreliable and/or the parts may be hard to obtain.

Therefore, the landlord may suggest that, subject to the agreement of the tenants, the items may be replaced if it assists the smooth running of the shopping centre (by having fewer escalator breakdowns) or the new air-conditioning may better cope with the fact that most/all office workers now have a computer on their desk, and the heat given off by such IT equipment cannot now keep the office at comfortable temperatures.

It is not unusual for tenants to actually request that the landlord should replace these items, knowing that they are not in disrepair and that there may be a number of years left before they cost more to repair and maintain than to replace.

Service charges are the major costs that businesses face which cannot easily be determined from year to year. Unless tenants have a cap on their service charges or have exclusions on their liability set out in their leases, they cannot predict what the costs will be from one year to another (unlike rents that only change at a known review date and business rates that change subject to the phasing rules).

Therefore, it is no surprise that disputes sometimes arise between landlords and tenants as to the sums which tenants are being asked to pay. Often, these disputes are about whether items such as lifts or air-conditioning systems can be repaired or whether they should be replaced. That type of dispute is where the tenants may be seeking a contribution from the landlord towards the costs of replacement of "big ticket" items of expenditure.

Another type of service charge dispute is not between the landlord and the tenants, but between the tenants themselves. Tenants may feel that other occupiers of an office building or shopping centre should pay a higher proportion than the proportion that the landlord has assessed them to bear.

When this occurs in an office building or a shopping centre, the landlord will usually look to the amount of floor space each tenant occupies and assess their service charge contributions based on that. However, the amount of floor area occupied by each tenant may not always produce a satisfactory solution to the amount of service charge each tenant should pay. An example of a type of property where this often will not be appropriate is that of a mixed use leisure scheme.

Mixed use schemes

Due to the Government's planning policies on building more homes, together with local authorities' aim for more housing has led to a rise of mixed use schemes in town centres, these type of schemes are becoming more common. As they do so, landlords whom would not normally become involved in residential properties, are having to deal with recovery of service charge sums from residential tenants. The concern for these landlords is that there is a strict regulatory regime in place that protects residential tenants which does not apply to commercial tenants. Additionally, the varying types of users may cause the landlord potential problems with apportioning the service charges sums.

Residential occupiers

There are a number of restrictions on what a landlord can recover from the residential tenants (which would take up a separate article), which include:

- Any works likely to cost £250 or more per flat, or a new contract that would cost £100 or more per year needs prior notification (known as section 20 consultation) to the residential occupiers, whether they are long leaseholders who have purchased their flats, or tenants on assured shorthold leases.
- Costs must be recovered from residential tenants within 18 months of the costs being incurred.
- Residents who are not satisfied with the management of the premises may apply to the leasehold valuation tribunal for the

managing agents to be removed and a replacement be ordered, even if it is against the will of the landlord.

This means that a landlord has to tread carefully when dealing with the residential occupiers of this type of property.

Section 20 consultation: recent case on invalid notice

A recent case that came before the Lands Tribunal illustrates the problems a landlord can face if they do not serve a valid section 20 notice on the tenants of a residential block or housing development where works are being planned.

Islington LBC v *Abdel-Malek* [2008] 1 P&CR DG4, heard at the Lands Tribunal in August 2007, concerned the issue of whether a notice served on the tenants complied with the requirements of section 20, as amended by section 151 of the Commonhold and Leasehold Reform Act 2002. It was held that the notice was not to be valid.

The local authority landlord wished to carry out repairs, "enhancements" and window replacements at a number of blocks of flats where the respondent held a flat on a 125-year lease.

The landlord tendered the works and received five responses from contractors. The landlord's section 20 notice to the respondent stated that her contribution would be a precise sum, which was just over £10,000.

The requirements of a section 20 notice states that at least two estimates for the works that are planned are passed to the residents. The definition of "the works" the Lands Tribunal was asked to consider and held was the works to be carried out to the respondent's particular block. This information was not sent to the respondent, in this case.

Another problem the landlord had in this case was that only a summary of the works was passed to the residents and not the copy of the estimate itself. The Lands Tribunal said that:

> the purpose of S.20 is to give a tenant sufficient information by way of copy estimates to be able to compare, and make observations on, the estimates for those works for which he is liable to contribute by way of a service charge.

As the notice served on the respondent was not valid, the only sum that the respondent, in this case, had to pay towards the circa £10,000 was £250: the maximum set in the Act and the regulations that support the statute.

The message to landlords of residential premises is clear: not serving a valid section 20 notice can cost you dear if you cannot recover the service charge monies expended on the premises as a result.

Commercial occupiers

Commercial tenants do not enjoy any statutory protection from the sums that may be charged. The best they can hope for is a cap on their service charge or stated exclusions from liability in their leases.

However, they may not be content with the landlord's apportionments of the service charge for the commercial tenants.

Due to the wide ranging uses on a mixed use leisure scheme, the floor areas of each use may not be acceptable to the occupiers of the estate. Some of the problems that landlords face are included in the following section.

Hours of opening

A night club which is only open 9pm to 2am may be identical in size to the next door unit, a café, which is trading from 7am for breakfast, right through until 11pm or 12pm. Clearly, floor area apportionments would not be appropriate here as the café is making use of many of the services such as cleaning, lighting, security, etc when the night club is not open. It would not be fair to ask each use to pay the same. A form of weighting of the floor areas of these uses may be acceptable to these types of tenants.

"Destination occupier"

There may be a use on the estate that is a magnet in attracting people to this particular mixed use leisure scheme, as opposed to others nearby. A type of use may be a cinema or a bowling alley, which tend to attract people to the cafes and restaurants nearby. While that magnet, or "destination occupier" is trading, the scheme may be seen as a success; however, if that occupier were to cease trading for any reason, it may have a detrimental effect on the other occupiers. They may feel that their weekly takings are adversely affected by the loss of that use and may ask the landlord to replace that use urgently and pay towards their service charges in the meantime. Whether the tenants would be successful in having the landlord pay towards the estate costs is one thing, but tenants may ask as service charges are just one of the other tenants' overall running costs.

Schemes with a wide variety of uses

Some schemes have such a wide variety of uses that even looking at the floor areas of the buildings as a starting point may not assist the landlord in assessing the correct apportionments of the service charges.

A mixed use leisure scheme that includes such uses (run by tenants in different ownerships) as, say, an outdoor running track, a bowling club, a golf course, a boating lake providing boat trips, etc fits into this category. As no use can be assessed by floor areas (most may only have a small building adjacent), other factors must be taken into account.

One method may be to look at the visitors to each use (a footfall count). Footfall methods of apportionments are not well liked by most leisure operators, or by the Royal Institution of Chartered Surveyors (RICS) Service Charge Code. However, they may prove a satisfactory starting point in order to apportion the costs. A use with, say, four times as many visitors to it than another should arguably pay a higher proportion of service charges than the use with the lower footfall.

Summary

There are no hard and fast methods of apportioning the sums that occupiers on a mixed use leisure scheme should pay, unlike office and retail schemes where floor areas have proved satisfactory to all parties. Landlords may have to look at all of the factors of the different uses and come up with a formula to put to the tenants to see if they would agree to that being the most appropriate in that scheme's circumstances. Additionally, if uses change over time, such as a destination occupier ceases trading, the position may have to be looked at afresh in order to assess the fairest way of recovering the running costs of the scheme.

19 Service Charge Disputes

Common disputes

Works to mechanical and electrical equipment close to lease expiry

Of all the likely potential areas for dispute between landlords and tenants, the most likely is usually works carried out to the plant close to the expiry of the tenants' leases.

The reason why this is so contentious is that it is these types of work to the building that tenants often consider is for other parties' benefit — yet paid by them — whether that be the landlord (who will have an improved building ready to re-let) and/or the next occupiers who will have had the previous occupiers pay for these items.

These works encompass what can be referred to as "big ticket" items and are one-off costs, ie they are not the usual service charge items such as cleaning and security which make up the bulk of an average service charge budget. These costs tend to only be incurred every 10 to 20 years and include:

- roof replacement
- air-conditioning replacement
- asbestos removal
- lifts replacement.

For whose benefit are these works: tenant or landlord?

Where a lease is coming to an end and a tenant is deciding whether to renew or not, a landlord will often wish to ensure that they carry out as much works as possible to the mechanical and electrical (M&E) equipment (where they maintain control over its maintenance) while there are tenants on leases that are obliged to pay for these works.

Clearly, if the commercial premises are let to a single occupier, it is very likely that that single occupier will have the responsibility to maintain the M&E equipment and other "big ticket" items, so the landlord can deal with the matter by arranging for a schedule of dilapidations to be served on the tenant shortly before the expiry of that lease or very soon afterwards. In that case, a landlord does not get involved with arranging for works to be done and seeking recovery of the costs through the service charge.

However, where the landlord is responsible for these one-off "big ticket items", they will wish to ensure that the works are carried out while they have tenants "on the hook" to pay towards them.

Lift replacement

In any commercial premises, whether offices, shopping centres and so on, it is usual for the landlord to have a programme of planned preventative maintenance. Therefore, it is likely that, after a number of years, the landlord would wish to replace the lifts, either because they are becoming unreliable or due to spare parts becoming harder to obtain, or both.

However, as has been seen from the cases referred to in previous chapters, the tenants are likely to resist these large one-off costs if they consider that the lifts are capable of repair and replacement of component parts and not the complete replacement of the entire lifts.

Asbestos removal

Similar to lifts replacement, this work involves a high level of costs to tenants and they are likely to object to having to pay for these works.

In addition to the tenants objecting to these costs, there is the additional issue that tenants may wish to quote: the fact that the current guidance from the Health & Safety Executive is not to remove asbestos but to leave it in place and have it carefully marked. So long as asbestos is labelled as such and is contained in an asbestos register held on site, the current view is that it should not be removed. Clearly, when a building is being demolished then there will be the need to remove it in accordance with the strict regulations in force, but while a property is in use, there should not be the need to remove asbestos if it is kept safe from being disturbed by labelling it.

Roof replacement

There have been a number of cases concerning roof replacement at the High Court and the Court of Appeal, mainly concerning whether a roof may be repaired or whether it needs replacing completely.

Without going through each case from the High Court and the Court of Appeal in turn, the general view is that it is the party who has the repairing responsibility that may reasonably determine the method by which works are undertaken. So, as seen elsewhere in this book, where the tenant has the responsibility for repairing a roof (or other items in a building) then the tenant may determine the way that it is carried out. Where it is the landlord (which is the usual way when recovering the sums via the service charge), then the landlord generally may determine the way in which the works are undertaken.

In relation to roof works, these mainly involve whether to repair the existing roof or to replace it with a new covering.

Clearly, repairing an existing roof is likely to cost less than replacing the same and a landlord who has a tenant obliged to meet the full costs of such works, without enjoying a service charge cap, is likely to wish to replace a roof than repair it. One of the reasons for this is that while the tenant's lease is running, the landlord has another party who has to pay for these works to the landlord's building: when a tenant's lease ends of a multi-let office or shopping centre, the landlord will have to pay towards those works, rather than the tenant. This leads to the landlord favouring replacement of items rather than repairing a roof from time to time.

Air-conditioning replacement

This is a very large cost for tenants to bear and unless it is in disrepair or has any defects, the case law from the *Fluor Daniel Properties Ltd* v *Shortlands Investments Ltd* [2001] EGCS 8 is likely to lead to tenants not having to pay towards these costs. Similar arguments that were referred to in earlier chapters of the book, in relation to dilapidations disputes, are put forward such as the equipment may be merely repaired at a lower cost than replacing the entire item.

There is no difference in the case law about whether this type of equipment may be repaired or replaced: so whether it is a service charge dispute or a dilapidations dispute, the same principles of repair versus renewal will be argued by advisors to the parties.

Who pays: landlord or tenant?

A landlord may notice some resistance from the tenants of a shopping centre or office building when the landlord has let it be known that they plan to carry out one or more of these "big ticket" items.

However, the landlord may decide to press on with the works, regardless of the objections or resistance received.

It is only when the tenant receives the additional costs in the service charge demands that they may consider whether to pay these sums or to consider withholding that part of the service charge request that relates to the disputed works. There are the additional considerations as to whether or not to withhold service charges reserved as rent and the landlord may instruct bailiffs or seek forfeiture of the lease for non-payment. However, it is a powerful tool for tenants to withhold such sums. The threat from tenants that they may not pay these sums which the landlord has spent can be enough for the landlord to consider whether or not they should proceed with these works. It would be a risky decision for a landlord to press on with such works, knowing that the tenants are objecting to them being carried out under the service charge.

However, it often happens that the works are carried out and the landlord does threaten bailiffs or forfeiture of the leases.

The right of a landlord to threaten, or even use, bailiffs will cease when the Tribunals, Courts and Enforcement Act 2007 comes into force. Although the Act has received Royal Ascent, the date when it will come into force has not been set; however, it is likely to be in early 2009 at the latest reckoning.

The Princes House Ltd *v* Distinctive Clubs *case*

Princes House Ltd v *Distinctive Clubs* [2007] 14 EG 104 (CS) involved the landlord being aware that works were required to be carried out to the roof of an office and retail building in Piccadilly, London W1.

The landlord was aware that the roof needed replacing and they had been carrying out patch repairs to it during the previous few years. The case came to the High Court and, subsequently, the Court of Appeal, after the tenant considered that the landlord had planned to delay the roof replacement works until a date when one of the tenant's cap on their service charge contribution had expired. This tenant, whom operated a casino in the basement of the building, argued that the works could have been done during the time that the tenant enjoyed a cap on its contributions to works to the common parts, which included the roof. If the works were carried out during that period, the landlord would have to bear that tenant's contribution to them and not the tenant.

Both the High Court and the Court of Appeal held that the works had been delayed and that the tenant should not have to pay their contribution to these works.

Turning again to the case of *Fluor Daniel Properties Ltd* v *Shortlands Investments Ltd* [2001] EGCS 8, the judge considered the position of which party should be paying for works when a tenant's lease is soon to expire and that tenant is being asked to pay for works that will be for the benefit of another party, particularly the landlord.

To quote from part of the judgment:

> In short the works — ie the standard to be adopted – must be such as the tenants given the length of their leases, could fairly be expected to pay for. The landlord cannot, because he has an interest in the matter, overlook the limited interest of the tenants who have to pay by carrying out works which are calculated to serve an interest extending beyond that of the tenants. If the landlord wished to carry out repairs which go beyond those for which the tenants, given their more limited interest, can be fairly be expected to pay, then, subject always to the terms of the lease or leases, the landlord must bear the additional cost himself.

It has often been said that landlords do wish to keep their buildings in good condition. Most tenants would not have a problem with that aim. The problem only comes about when a landlord wishes to carry out works to the plant and equipment during the end of a tenant's lease. It is that timing matter that leads to tenants thinking that they are paying towards items that they, themselves, will not get the benefit from.

Landlords like to carry out works while their tenants are liable to contribute towards the costs and not afterwards. This leads to the suspicion that items may be replaced or renewed even though the items are not in disrepair.

To quote Jonathan Karas QC of Wilberforce Chambers:

> It is plain that an item is not in disrepair simply because it is old. Thus, if an item of plant reaches its life-expectancy in accordance with CIBSE tables, this will not in itself mean that the item is in 'disrepair'; likewise if the item is in working order but obsolete it is not in disrepair. So, if an item of plant is new at the start of a 25 year commercial lease and has only a 20 year life-expectancy under CIBSE or other tables, this will not in itself entitle to the landlord to seek to recover the cost of replacing that item shortly before the expiry of the lease as a cost of 'repairing' that item.

So it can be seen that if an item is still carrying out the function that it is designed to do, it should not be replaced just because it has exceeded a certain life span. If it is, a tenant may bring proceedings against the landlord for trying to recover monies for works when that item could have been repaired instead.

Improvements in service charge disputes

The difference between the position of improvements in a dilapidations claim and a service charge dispute is that in the former, the landlord is unlikely to receive the monies for a disputed dilapidations item unless and until a settlement is reached on all of the items in the schedule of dilapidations. However, with a service charge dispute, often the landlord will have gone ahead and done the works and charged it to the service charge and the dispute would only commence when the tenant sees the costs in the service charge request for payment.

This can lead, therefore, to items in a service charge being strongly contested, where a tenant considers that they are being asked to pay for items that improve the landlord's property where the landlord has gone ahead and done these works.

Provided that the landlord has carried out the works to be recovered in the service charge before the expiry of the tenant's lease, the tenant is likely to have to pay towards them. Assuming that the works have been done during the currency of the lease, the tenant may object to them on the basis that they are improving the building, when the lease only permits maintenance and repair of the items listed in the schedule of service charge items.

Chapter 17 covers the position whereby, in certain circumstances, tenants are able to argue that a landlord is, in fact, trying to improve their building at the tenant's cost; however, the general rule shown below applies to maintenance, repair, renewal and replacement of items that are recoverable by a service charge:

1. Maintenance means the work of keeping something in proper condition; the upkeep of a property or equipment that services a property. Maintenance has also been interpreted by the courts to impose and obligation of wider application than an obligation to repair.
2. Repair means to restore to a good or sound condition after decay or damage; to mend a property or equipment that services a property.
3. Renewal means the renewing of part of something that has become in a state of disrepair or decay; especially mechanical and electrical (M&E) equipment servicing a property.
4. Replacement means the act of furnishing one thing for another that is equivalent to that changed.

So, it can be seen that whereas on the face of it, maintenance and repair appear to mean a similar obligation yet, in reality, they are not the same and maintenance is a more onerous obligation than repair for a tenant to covenant in a lease. Similarly, renewal covers instances where a lower degree of expenditure is generally incurred than replacement, with the latter meaning that the whole item is changed, not just the component parts.

It is interesting to note that commercial property leases would very rarely permit the landlord to recover items of improvement in the service charges; this often appears in leases of residential premises. A long lease of a flat or maisonette often contains wording that permits the landlord to recover items of expenditure that were incurred with the maintenance, repair and improvement of the common parts.

Sinking and reserve funds

A sinking fund is the name given to monies that are collected from tenants on a regular basis (usually quarterly, but could be annually) in order to build up a sum required to pay towards large "one-off" costs that may be required during the lease term. For example, a landlord may plan to replace the lifts in a building in five years time and, in order to ensure that the tenants are not hit with a large request for monies in that fifth year, the tenant(s) is/are asked to pay towards a sinking fund in order to build up the likely sum required in order to carry out and pay for these works.

A reserve fund is designed to equalise the fluctuations that may occur with regularly reoccurring service charge items payable each year. These are not that popular with occupiers as the landlord is not obliged to hold such funds as a provision against future large expenditure on replacement of items.

The monies are held in an interest-bearing account and are used whenever major replacement works are required or unforeseen costs are required to be spent.

Disputes arise between parties in relation to these monies that are being collected, usually on the basis that too high a sum is being collected by the landlord to pay for the likely works required, or the landlord does not wish to return the monies in the fund to the tenant or tenants at lease end.

Depending on the precise lease wording, the landlord may be able to retain those monies or may be obliged to return the same to the tenant(s) at lease end.

In *Secretary of State for the Environment* v *Possfund (North West)* [1997] 2 EGLR 56, the tenant was required to pay towards items including the replacement of the air-conditioning system. At lease end the monies paid by the tenant towards the costs of this had not be spent and the landlord was asked to return those monies by the tenant. The lease was silent on the return of those monies and the landlord resisted such return of monies.

The case came to court and the judge ruled that, under the terms of that lease, as soon as the monies were paid over by the tenant, they became the landlord's absolute property. The lease did not contemplate a position whereby those monies could be returned to the tenant. Therefore, the landlord was able to retain those monies and carry out the works for which they were designed after lease end.

However, the opposite view came out in the case of *Southwark Roman Catholic Diocesan Corporation* v *Brown's Operating System Services Ltd* [2007] EWCA Civ 164. The Court of Appeal decided that unspent monies could be returned to the tenant at lease end, notwithstanding the fact that the lease did not expressly state this.

The position therefore is dependant on what the leases state should be done with the monies and if that is not clearly stated, then the circumstances from other sections of the lease that may indicate the intentions of the parties when drawing up the lease.

20 Avoiding Dilapidations and Service Charge Disputes

Pre-lease measures

The best way for a tenant to mitigate their dilapidations position under a lease is to avoid the obligations from arising in the first place.

Probably the most effective way of avoiding dilapidations liabilities is for prospective landlords and tenants to take professional advice from surveyors and solicitors with experience in landlord and tenants covenants and how they relate to dilapidations and service charges when negotiating a lease. The involvement of experienced advisors at this early stage will allow the parties to properly consider the lease intentions and obligations.

A protective schedule of condition

From a tenant's perspective, it is favourable to seek to have the repair, decoration and maintenance related obligations on a lease all independently limited by way of express reference to a schedule of condition of the property. Whether a landlord accepts this is often a commercial decision based on the strength of the market.

If agreed, the schedule should be recorded at the commencement of the lease, and signed by landlord and tenant parties and attached to the lease. It is also advisable to have a comprehensive photographic record of the property recorded at the same time and included within the schedule of condition and the lease.

If the tenant is seeking to limit their future liabilities, then they should endeavour to have any reference to a schedule of condition drafted so that their future obligations require them to put it in "no better" condition, or similar legally binding text.

Inventories of fixtures, fittings and finishes, etc

The risk and degree of future dilapidations and service charge claims can also be anticipated and better managed during letting negotiations by the inclusion within a lease of detailed inventories that better describe aspects such as the extent and nature of the landlord's fixtures, fittings, finishes, plant and services present at the commencement of the term.

The inventories could be supplemented by clear as-built floor, "reflected" ceiling and site plans that show the configuration of the demise at the commencement of the lease term. These can be prepared unilaterally as a last resort.

Such inventories are not current common practice but where they do exist, they typically assist both landlord and tenant surveyors retained on dilapidation matters by providing a clear and detailed point of reference to take into consideration when a claim is prepared.

The merit of commissioning detailed inventories will be dependant on the size and nature of the demise and the risk and value of future dilapidations and service charge claims. Where the prospect for a sizeable claim exists, the costs incurred in producing the documents may well pay dividends in the long term by allowing accurate claims to be prepared that reduce dispute attendance and avoidable litigation costs.

Revising lease covenant terms

An alternative to having a schedule of condition attached to the lease is to seek to vary or revise the drafting/phrasing of the tenant's covenants of the lease. The future obligations to be imposed by the tenant's covenants for repair, decoration and maintenance obligations, etc should be kept to a reasonable and sensible level according to the nature of the property, tenancy, use and length of the proposed lease.

For example, a tenant looking to take a short term commercial lease for, say, three years would be somewhat foolhardy to accept widely drafted repair obligations which includes requirements for upholding, rebuilding or supporting the property or that may require the renewal of carpeting at the end of the term. If the tenant's position cannot be protected by the introduction of a cross-referenced limiting schedule of condition, then consideration should be given to having the future repair and decoration obligations limited by, say, excluding future deterioration arising from "fair wear and tear" or an inherent defect.

Better lease covenant drafting

There are many extensive books on dilapidations case law that indicate the sheer extent of dilapidations claims that end up in litigation. The property press publish regular articles and reports on dilapidation cases and developments on an almost weekly basis and there appears to be no sign of this litigation ceasing.

When judgments are read in detail, the vast majority of cases will centre on some relatively simple disputed issues where the disagreement between the parties will often be a result of an ambiguously or poorly drafted lease clause that is open to too wide an interpretation. Often, due to cost consequences for a claim, the two opposing landlord and tenant parties will seek to polarise their positions at opposite ends of the claim spectrum, so that "never the twain shall meet". Inevitably therefore, the dispute gets referred to the court.

Clearer drafting of leases will help avoid these types of disputes from occurring.

Commissioning a building survey

Where a schedule of condition or other pre-emptive protective measure is not permitted for attachment or inclusion within the lease, then the prospective tenant should obtain a building surveyor's report prior to completing the lease. This will enable the tenant to give due consideration to any issue or risk identified within the report so they may seek to manage or omit the risks identified.

Where the landlord is unwilling to consider a tenant's requests, then at least the tenant can make an informed decision on whether or not to proceed with the lease; or whether to look at alternative letting options.

Encouraging tenants to take advice

Despite the various measures available to a tenant to avoid or reduce future dilapidations and service charge disputes during letting negotiations, experience shows that tenants are generally reluctant to incur expenditure on professional advice before they commit to a lease.

Tenants all too frequently do not appreciate the significance of the covenants of a lease until it's too late and only regret their failure to have taken professional advice at an earlier stage. The tenants alone are not solely responsible for these prevailing attitudes and the Government has made it clear that it will legislate where necessary if the professions involved fail to address the issues and better protect the layman tenant.

In the face of the Government's criticism, there are efforts being made by the various professional bodies involved in commercial lettings to better educate and inform prospective tenants of the risks involved with taking a lease. The *Code for Leasing Business Premises in England and Wales 2007* (3rd ed) contains many sensible measures to be considered by landlords and tenants when granting leases for leasing business premises and is endorsed by a number of major professional bodies including the Government, the Law Society and the Royal Institute of Chartered Surveyors (RICS).

Arguably though, one of the barriers that have stood in the way of a more commonplace consideration of dilapidations during lease negotiations is the lack of data available to surveyors, agents, solicitors and the public generally on the value of the dilapidations sector and average claims data according to size, type and use of leasehold properties.

If the dilapidations claims sector and average claim data were better understood, then tenants could use the average data to quickly appraise a reasonably likely future dilapidations claim value/risk before committing to a lease. In all probability, when the value/risk of a average potential future claim is understood, the prospective tenant would then consider the expenditure on early pre-lease advice and professional services as being essential in mitigating future costs.

The good news is that plans are in place for dilapidations market value studies and perhaps once market value data is available, agents, solicitors and surveyors alike will be more informed and better placed to provide suitable pre-lease guidance to their clients. This is turn would encourage landlords and tenants to take the proactive dispute avoidance measures, outlined above, prior to committing to a lease.

Lease term proactive measures

Regular scheduled joint inspections/meetings

Regular site meetings can be offered by either the landlord or tenant during the lease term and can be tremendously beneficial. Whether initiated by landlord or tenant, these meetings can:

- raise issues that might ordinarily not be ascertained until after lease end
- avoid the formality and cost of interim schedules
- promote sensible dialogue between the landlord and tenant
- result in both landlord and tenant expectations being correctly managed in respect of the lease end dilapidations claim
- alert one party to the other's breach of covenant, thereby facilitating actions to mitigate the damage to the demise. This could involve unilateral actions such as the preparation of an interim schedule, or could be a combined approach to the benefit of both the landlord and tenant.

Such suggestions of a meeting should be put politely and in writing as evidence of the party's conduct during the term of the lease. Experience of this approach suggests that it promotes dialogue and reduces the potential for acrimony at lease end.

Planned maintenance programmes and budgeting

Whether tenants occupy the property on internal repairing only (IRI) or fully repairing and insuring (FRI) terms, they should adopt a rigorously planned property maintenance (PPM) programme during the lease to identify, prioritise and cost the required repairs as a matter of good practice.

If this PPM procedure is tied in with the lease end dilapidations liability, closely referenced to the lease ad licences, the result will be that the terminal dilapidations liability will be clearer and, in many cases, reduced for the tenant. For example, there is no point in a tenant decorating the demise 15 months before lease end when the lease requires them to decorate "during the last twelve months of the term". Savings in respect of costs and time resulting from such a holistic approach to the demise will pay direct financial dividends.

The recognition during the term of the eventual dilapidations liability will allow sensible budgeting and increase the potential for a negotiated settlement in the event the liability is ascertained in sufficient time to allow the works to be undertaken. If the tenant has advanced notice of the likely impact on their cash flow they will be more able to deal with the liability in a proactive manner.

FRI versus IRI

The demarcation of responsibility for repairing the structure of the building is normally addressed within the lease within the sections of the lease that cover tenant repairing covenants and landlord repairing covenants respectively. In theory, there should be no overlap between the areas of the building that are landlord and tenant responsibility respectively. In practice, a lease that is drafted by a lawyer who might not have visited the premises, and often without the assistance of a surveyor, can result in confusion.

Where a lease is FRI, it is probable that the tenant will have responsibility for all elements of the building, although the precise terms should be checked. Where a lease requires that the tenant is only responsible for internal repairing, the clarification of responsibility can be problematic. If this ambiguity is not addressed during the lease term, it is likely that confusion will arise at the end of the lease when dilapidations are discussed.

If there is confusion between whether an item of the building is the responsibility of the landlord or tenant, the following approach should be adopted:

- Look at the lease covenants. The express covenants may explicitly explain the responsibility of each party to the lease in respect of repair.
- Look at the definitions within the lease, and relate them to the tenant covenants, landlord covenants and the service charge clauses.
- Establish who has actually maintained the contentious elements of the building. Established patterns of behaviour in relation to the repair of a building can create binding legal obligations. For example, where a landlord repairs parts of the premises that are actually demised and recovers the costs through the service charge, this can reduce the tenant's obligation to deal with these

particular matters at lease end. Legal advice should be obtained if such an argument is pursued and becomes contentious.

Lease end dispute management

Joint expert appointments

Jointly appointed expert surveyors dealing with dilapidations claims are extremely rare. The principal reason for this is that the exact role of the surveyor at the end of the lease is unclear. In theory, the expert should be appointed by both landlord and tenant and would be asked to independently consider the lease end dilapidations liability.

The jointly appointed independent expert surveyor and/or a jointly appointed lawyer could therefore act in a quasi judicial capacity from the very start of a dispute. This works in other dispute areas at the pre-litigation stage and is a key principle of the Civil Procedure Rules (CPR) 1998.

While the Property Litigation Association's (PLA) *Dilapidations Protocol* does not formalise the role of the surveyor as an expert, the CPR *Default Protocol* encourages jointly appointed independent experts to resolve disputes at the pre-litigation stage. Section 8.3 of the Default Protocol (May 2008 ed) states:

> If the assistance of an expert is needed then the parties must, wherever possible, try to save expense by appointing only one expert.

If either party considers that the approach would be likely to promote the reasonable settlement of the issues in dispute, a letter should be issued to the opposing party explaining the basis of the suggestion.

Better use of ADR

Alternative dispute resolution (ADR) was a cornerstone of the CPR when they were first introduced in 1998. The concept was key to Lord Woolf's aim of reducing the amount of litigation.

The options embodied within the concept of ADR include:

- Discussion and negotiation.
- Mediation (a form of negotiation with the assistance of an independent third party).

- Early neutral evaluation where an independent party (often a lawyer or an expert) would give an opinion on the merits of a dispute or suggests a solution.
- Arbitration where an independent third party makes a binding decision.

Joint site meetings

Joint site meetings with both landlord and tenant surveyors at the start of the claim, in order to prepare a joint and agreed schedule, can expedite the resolution of a dispute and even avoid disputes entirely. This is used in the Housing Disrepair Protocol and may avoid the concerns of using just one "joint" expert surveyor where each party might wish to retain their own expert.

Cohesive professional teams

Because dilapidations claims have been traditionally left to building surveyors to consider, prepare, serve, negotiate and settle, many claims have become mired in avoidable disputes simply because the building surveyor may not necessarily possess the appropriate legal or technical knowledge on certain aspects of the claim process.

Any surveyor appointed on a dilapidations claim should not seek to work in isolation but should maintain regular contact and liaison with the client's other advisors, including solicitors, mechanical and electrical engineers, structural engineers, valuers, agents, etc.

The driving force behind the legal processes and tactics should be the client's solicitor. However, the co-ordination of the technical surveying, schedule documentation and quantification of any claim should be undertaken by the building surveyor.

Regular meetings throughout the life of a dispute may prove invaluable and the professional team, in conjunction with their client, should strive to work together and should be prepared to critically appraise the various constituent parts, services and advice that make up the claim.

In many cases, the prospects of success of a claim or the defence will be determined by the effectiveness and cohesiveness of the professional teams. Parties supported by an effective team will normally manage to avoid potentially costly errors or futile disputes and will often adopt a more realistic, fairer and more considered approach to proceedings and conduct throughout. Effective team working is therefore essential if many of the common disputes incurred during dilapidations proceedings are to be avoided.

Better professional education

If the dilapidations community is viewed as a whole, then perhaps the best way of avoiding disputes in future is for the professional bodies, associations and forums with a common interest in dilapidations to work more closely together. The interested bodies should seek to pull resources, work together and to put into place education programmes for their members that better explain the concept and principles of dilapidations and the legal processes involved.

Through better education, the professionals will be able to educate, inform and guide their landlord and tenant clients on the processes and principles. Where the clients receive clear fair and impartial guidance, it then becomes easier to manage client expectations as to the likely fair settlement is for any given claim.

Further reform

Statutory change as a solution for the future?

Despite the progress over the last 750 or so years, damages claims and the law on waste, contractual damages, landlord and tenant remain a point of concern and debate.

Lessons could be learnt from other protocols that have radically changed other areas of civil dispute, eg the Housing Disrepair Protocol and the Personal Injury Protocol. Solutions that should be debated more widely amongst lawyers, surveyors, landlords and tenants include a statute specifically addressing the common problems of dilapidations.

This approach could completely restructure the dilapidations dispute resolution process, in the same way that the Party Wall etc Act 1996 regularised the disputes between owners of adjoining property. Such a procedure could include the creation of the role of a "third surveyor" for dilapidations disputes to potentially oversee disputes on items within a claim as they arose, at zero cost unless consulted. Such a procedure could also be enshrined within leases and would automatically create a more reasonable climate.

It seems that common law and statutory principles governing dilapidations claims continue all too often to be misunderstood, abused or simply ignored. There is a growing view that in many cases, the law and processes are being so wilfully or recklessly abused that many dilapidations claims are verging on fraud.

In the current climate, the Government are actively looking at further reform and professional bodies such as the RICS, The Law Society, The RICS Dilapidations Forum, The Property Bar Association and the Property Litigation Association are also looking to promote best practice and better educate their members. However, this comes with mixed results and sometimes conflicting approaches, further muddying the waters.

Most recently in 2006, the Law Commission Report (no 303) reported on the current law on termination of tenancies for tenant default. The draft bill within that report contained wide ranging proposed reforms that would also be of significant relevance to dilapidations proceedings.

Future predictions

The fact that a protocol for dilapidations has not been adopted creates an unsatisfactory state of limbo for commercial landlords and tenants, their surveyors and lawyers. This situation allows the current problems to persist, creating a poor public and industry perception of surveyors who are too often viewed as opportunistic advisors or negotiators, rather than experts in their field. Continuing problems in some instances include:

1. Highly adversarial negotiations, with surveyors dealing with dilapidations disputes initially adopting much polarised positions.
2. Exaggerated claims, with many landlord claims halved by the time they are settled.
3. Poorly set out claims, where senior landlord surveyors serve and attempt to negotiate claims drafted by a junior member of the team, often resulting in disjointed and poorly set out claims.

Lessons should be learnt from other protocols that have radically changed other areas of civil dispute, eg the Housing Disrepair Protocol and the Personal Injury Protocol. Solutions that should be debated more widely amongst lawyers, surveyors, landlords and tenants include:

1. Joint appointments: with a jointly appointed independent expert surveyor and/or a jointly appointed lawyer, either of whom could act in a quasi judicial capacity from the very start of a dispute. This works in other dispute areas at the pre-litigation stage and is a key principle of the CPR.
2. Joint site meetings with both landlord and tenant surveyors at the start of the claim, in order to prepare a joint and agreed schedule where possible. This is used in the Housing Disrepair Protocol and may avoid the concerns of using just one "joint" expert surveyor.
3. The schedule is a document served in contemplation of litigation; surveyors could have CPR "expert" status from the time a schedule is served. This would require a CPR compliant "statement of truth" to be signed by the surveyors or their landlord/tenant clients. By utilising the standard wording for a statement of truth set out in the Practice Direction to CPR Part 35, perhaps the widespread concern amongst surveyors regarding the suggestion that they be required to sign a written endorsement of the landlord's loss (ie PLA Protocol section 4.8.3–4) could be avoided. A fair and equal obligation could also be placed on the tenant's surveyor when responding with their counter schedule.
4. Utilising a "third surveyor", similar to the Party Wall etc Act 1996. The mere presence of such a third surveyor (at zero cost unless consulted) could restrain the initial positions of the landlord and tenant surveyors.
5. Formalisation of the dilapidations resolution process within the lease, on a similar basis to commercial lease rent review clauses.

Whether such changes to the protocol were incorporated, would require extensive national consultation between: surveyors; lawyers; the Ministry of Justice; landlords; and tenants, but might help to persuade the Ministry of Justice that everyone involved in commercial dilapidations disputes had bought into a suitable protocol.

Having reviewed the history and current state of the law on dilapidations, further changes are inevitable. Given the current perception on the inadequacies and complexities of dilapidations proceedings, how long will it be before the dilapidations claims process itself is specifically reformed and regularised, with possibly alternative dispute resolution procedures imposed in a "Dilapidations Act"?

Index